Where Are We Heading?

Foundational Questions in Science

At its deepest level, science becomes nearly indistinguishable from philosophy. The most fundamental scientific questions address the ultimate nature of the world. Foundational Questions in Science, jointly published by Templeton Press and Yale University Press, invites prominent scientists to ask these questions, describe our current best approaches to the answers, and tell us where such answers may lead: the new realities they point to and the further questions they compel us to ask. Intended for interested lay readers, students, and young scientists, these short volumes show how science approaches the mysteries of the world around us and offer readers a chance to explore the implications at the profoundest and most exciting levels.

Where Are We Heading?

*The Evolution of Humans
and Things*

Ian Hodder

Yale UNIVERSITY PRESS
NEW HAVEN AND LONDON

Templeton Press

Yale University Press books may be purchased in quantity for educational,
business, or promotional use. For information, please e-mail
sales.press@yale.edu (U.S. office) or sales@yaleup.co.uk (U.K. office).

Designed and set in Hoefler Text by Gopa & Ted2, Inc.

Printed in the United States of America.

Library of Congress Control Number: 2018939537
ISBN 978-0-300-20409-4 (hardcover : alk. paper)

A catalogue record for this book is available from the British Library.

This paper meets the requirements of ANSI/NISO z39.48-1992
(Permanence of Paper).

10 9 8 7 6 5 4 3 2 1

For Lynn

Contents

Preface

If you wish to make an apple pie truly from scratch,
you must first invent the universe.
—CARL SAGAN (ASTRONOMER AND WRITER
[1934–1996]), COSMOS

THIS BOOK deals with a problem that has bedeviled archaeologists, anthropologists, social theorists, philosophers,
and evolutionary biologists: Does human evolution have
a direction? If so, why? These questions dominated debate in
the nineteenth century and in the mid-twentieth century, but
consensus today is that biological evolutionary development is
nondirectional. The same holds for human cultural development:
Humans, their cultures, and their societies change or evolve, but
they are not thought to evolve in any overall direction. One reason for this view is that directional arguments lead to the notion
that some societies or cultures are more advanced than others.
Because such theories were found to be repellent, there were long
periods in which evolution was rejected in favor of diffusion as a
motor for change.

But we have an urgent need to reconsider the assumption that cultural and social evolution are nondirectional. There is clear archaeological evidence for an overall trend in human evolution toward greater dependence on and entanglement with things. As a species we keep producing more and more stuff. This brings significant changes in many areas of life, but our dependence on an increasing mass of things has ravaged the world in which we live, leading to global problems such as possibly irreversible climate change. We need some insight into why our species is heading in this direction.

I should make absolutely clear that I am not concerned in this book with arguments for intelligent design. This book is not about invisible hands or divine destinies. I am interested—in this book—only in tangible evidence and the practical intersections of humans and things that have led to long-term change.

My argument could be called evolutionary. What might this mean? Two main types of evolutionary theory have dominated science over the last two centuries. In the nineteenth century, social evolutionists argued that societies progressed toward advanced urban industrial systems; related arguments emerged again in the mid-twentieth century and are today phrased in terms of movement toward complexity. Such arguments are directional but also teleological in the sense that societies are seen as moving toward some end or goal. The problem is that the end (the telos of advanced complexity) is also the cause of change (humans become more complex because they want to, or human societies become more complex because it is the nature of complex systems to become more complex). This approach does not explain very much.

The other main type of evolutionary theory is inspired by Darwin, and over recent decades it has come to play an important role in accounts of long-term social and cultural change, sometimes in combination with complexity theory. In the latter case, it is sometimes argued that more complex systems get selected for, because they allow better adaptations to the environment.[1] Darwinian approaches usually eschew directional change; rather, change comes about through variation, selection, and transmission. There is no reason that societies should move in one overall direction; they should just adapt to changing environments.

The trouble with social evolutionary theories is that they are teleological. The trouble with Darwinian theories is that they do not explain the archaeological evidence for overall directional change. This book argues that it is possible to find a third way; it is possible to build a theory of directional change without returning to the dangerous teleologies of nineteenth-century thought. We could call this approach "evolutionary," but I have largely avoided using that term in the text of this book because it has come to be closely associated with biological evolution. While I integrate biological evolution into my notion of entanglement in this book, I do not wish to reduce entanglement and long-term change in entanglement to biological evolution or to the metaphor of biological evolution.

Many archaeologists and sociocultural anthropologists dislike the word "evolution" when it is applied to their field. Some of this dislike concerns the idea of progress toward "higher"-level societies, while some of it concerns the lawlike, reductionist nature of some evolutionary arguments. It would be nice to be able to use other terms, such as "development," but these also have their problems. My main reason for using the term "evolution" at all in this book is that I wish to focus on the sense of "unrolling" that lies in the word's origin. Long before Darwin, it was used to refer to the unfolding of events. In arguing for a directionality in human evolution, I am using the word in its sense of "unrolling," as well as invoking the idea of gradual change. But I also wish to rescue the term from its associations with progress and law-like development. Another aspect of its etymology refers to how things "turn out"—hence to the contingency of developmental change. In this book I propose to redefine human evolution as directional (path-dependent) but also contingent and complex (nonreductive).

My goal is to find a third way between biological and sociocultural evolution by focusing on "things." Following much current research in cognitive evolution, materiality studies, and archaeology, I argue that human "being" is thoroughly dependent on made things. Since the first tool, humans have always dealt with problems by changing things.[2] This dependence on things has produced our "humanity," but it has also entrapped us in yet greater dependence. Things are unstable: They have their own processes that entangle humans into their care. For example, humans domesticated wheat, came to depend on it, and still do today. But wheat, once domesticated, can no longer readily reproduce itself. So dependence on wheat has entangled humans into the increased labor of plowing, sewing, weeding, harvesting, and processing. We increasingly get caught in a double bind—depending on things that come to depend on us, so that we must labor and develop new technologies.

As humans found solutions to the demands made by wheat, they became more and more entangled in inventing new technologies, creating yet more things. And the things kept interacting with each other and generating new demands. Today these interactions have created problems on such a scale that continued technological response seems difficult. We struggle to find and institute solutions to global warming and environmental deterioration.

The examples I describe in this book range from the introduction of fire, wheat harvesting methods, the wheel, cotton, and opium to dams and Christmas tree lights—an assortment that allows me to make different points about human-thing entanglement without losing sight of key similarities. All of these things have brought major changes in lifestyle and in what it means to be human, but they have also trapped humans into greater entanglements and still greater innovation.

Using this strange assortment of examples, my approach is unashamedly archaeological. By this I mean that I deal with the very long term and am concerned with everyday material practices. I start with archaeological evidence for long-term change over the broad sweeps of time unavailable to other disciplines dealing with the products of human labor, and I finish by arguing for an "archaeological" understanding of many of the problems our species faces today. The grand view that is possible only through an archaeological lens has much to contribute to fundamental debates about what it is to be human and about where we might go from here.

There is increasing questioning of the notion that growth in gross domestic product (GDP) is a necessary condition of contemporary capitalism. Economists have shown very close links between economic production, material consumption, and carbon emissions. Several economists also argue that economic growth is a recent phenomenon and that it is in no sense "natural." There are, of course, the new consumerism and ethical shopping movements that, among other things, aim to tread more lightly and manage our relationships with things more carefully. Some economists have increasingly argued for capitalism without growth. They have identified the reasons for the current preoccupation with growth as, for example, the need to limit unemployment and increase income. And they have identified a range of factors that have prevented low or no growth in the face of environmental deterioration: In some cases, environmental limits have already been reached, emerging economies are having increased impact; there is social inertia; easily useable resources have already been used up; and the absolute limits of increased efficiency (for example, in crop yields) are being approached.[3]

Important as these discussions are, there is a danger that they remain myopic. They focus only on the short term and on recent historical times and so do not tackle more fundamental causes of our commitment to growth and our inability to respond adequately to the environmental problems caused. In this book I look over the long term, with an archaeological eye. From such a perspective, one finds gradually increasing human-material entanglement right from the start, millions of years ago. There have been ups and downs, stops and starts in the gradual accumulation of material stuff, but over the long term a steady exponential increase. Growth is not a new thing; it has great temporal depth. It seems likely, then, that the causes of growth are much deeper and more complex than provided by historians and economists of the recent past. In this book I argue that increasing material stuff is a part of human experience, not because of some innate human desire to have more, but because of the way we are entangled with the world. This book does not offer solutions to the current impasse between growth and environment, but it does argue that a long-term "archaeological" view offers a different and important perspective.

Acknowledgments

I WISH TO THANK all those who commented on earlier drafts of this book, including two anonymous reviewers. I especially wish to thank Bill Frucht for the most diligent combing of my text that I have ever come across, as well as Susan Arellano, Paul Wason, and the John Templeton Foundation. Lindsay Der, Scott Haddow and Katy Killackey assisted with editing and formatting, permissions and illustrations. I am grateful to Ian Morris for figures 1.1 and 1.2 and to Bright Zhou for discussions about the biological in entanglements, leading to figure 3.1. I am grateful to Dorian Fuller for figure 4.3 and for discussions about plant domestication. The book was written while at the Maison des Sciences de l'Homme, and I am grateful to Alain Schnapp and Jean-Luc Lory for their help in Paris; while I was a Director's Guest in Humanities at the American Academy in Rome, where convivial support was provided by Kimberley Bowes; while a Visiting Fellow at Keble College, Oxford, thanks to Chris Gosden and Lambros Malafouris. I am most of all indebted to Lynn Meskell for her comments and more generally for the many years of discussion and dialogue on which the book is based. Her wisdom, wit, stimulation, and companionship have made this book possible. This book is dedicated to her in unbridled admiration and gratitude.

Where Are We Heading?

The Question

THE EVIDENCE for an overall direction in human development seems clear. Most of us would agree that humans as a species have progressed, even if we identify different markers of progress. Some might point to advances in health and medicine, others to great architectural and engineering feats, others to artistic achievements, communication, media and ease of travel, libraries of knowledge, stable institutions, governments, bureaucracies, democracies. In a series of publications, Ian Morris has used a range of measures to document social development in Europe and Asia since the Pleistocene.[1] (Some examples are found in figures 1.1, 1.2.) It is of course infamously difficult to measure social development over time, and many social scientists balk at their colleagues' efforts to reduce the fine contextual variation of human life ways—all the differences in art, culture, and moral purpose—to some simplified quantitative measure.

While acknowledging those reservations, Morris suggests four ways of measuring social development. The first is energy capture: the amount of energy a society can extract from the environment in order to further its aims. I return to this notion in chapter 2, but for the moment I note that it is very difficult to measure energy capture in early prehistoric societies. Morris uses a variety of archaeological and other sources to estimate the number of calories used per person per day for food and for other purposes. The second measure is organizational complexity, and here Morris uses the proxy of urbanism. The size of a society's biggest city gives an indication of how resources are organized to support and manage large concentrations of people. The third and fourth measures are information technology and the capacity for war making.

Figure 1.1 shows how these measures have increased over the past fourteen millennia in the most developed core areas in the

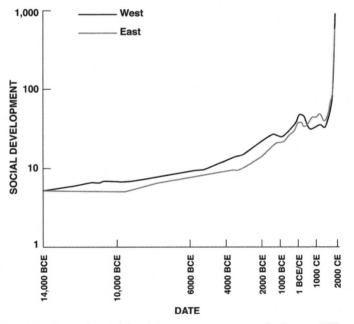

Figure 1.1. Overall social development in societies in the East and West. Source: Ian Morris. Used with permission.

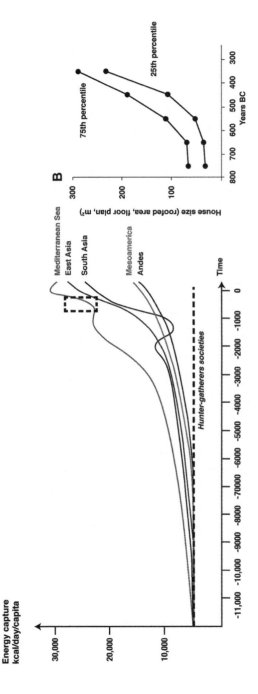

Figure 1.2. Evolution of Energy Capture since the Neolithic Revolution. (A) Evolution of energy capture during the period represented in B. (B) Evolution of Greek house size, a proxy for energy capture. Source: Ian Morris. Used with permission.

West (western Europe and North America), and in the East (China and Japan). Even if the details of the ups and downs are incorrect, the overall trend is clear. The vertical axis in the graph is logarithmic, so the increase is even more marked than appears. The rate of change has been exponential.

But why concentrate just on two areas in the world? More recently, Morris has charted the growth of energy capture in the five most intensive agricultural zones from 11,000 BCE to 1 CE (figure 1.2).[2] He used variables such as house size, population density, and the main city's population as proxies for energy capture. The charts hide much local variability. Locally there is much evidence of "booms and busts," with different areas seeing "booms" at different times.[3] Morris focuses on the boom areas and shows that in these areas there is an overall increase in energy capture through time.

But what of the many parts of the world not accounted for in Morris's charts? In some areas, such as precolonial Australia, overall trends may appear difficult to discern. And yet a recent review of the prehistory of Australia tells of continual change since the continent was first settled around seventy-four thousand years ago.[4] After 18,000 BP there were new technologies (such as the boomerang), more diverse and efficient stone tools, new migrations, new forms of art, increased territoriality linked to increased population, more village formation, and more extensive trade. Even parts of the world that seem, by some measures, as least developed have seen long-term increases in energy capture and organizational complexity, even if at a different rate.

Is this an overall upward trend in which all societies have participated equally, or have some societies massively increased at the expense of others? The latter seems best to fit the evidence. The degree of variation in energy capture in the world today is striking. The World Bank estimated that the total annual energy consumption per capita in 2011, measured in kilograms of oil equivalent, was 7,032 in the United States but only 205 in Bangladesh. NationMaster estimates that the annual electricity consumption in Iceland was 31,147 kilowatt-hours per capita in 2006, while in Chad it was 9.41 and in the Gaza Strip 0.167. Thus, while some societies have bumped along at or returned to a rate of increase close to zero, others have shown staggering increases. Any average value represents some range, but the lines in Morris's graphs hide stark inequalities.

Even if we are using them only to show rough trends, numerous difficulties remain with these types of measures. For example, there are many kinds of organizational complexity. Australian Aboriginal society is famously complex in terms of its languages and kinship systems, whereas New York has a baffling subway system. We often hear that more complex societies are more specialized with more levels of integration and organization, but this depends a great deal on what aspects of organization we are studying.

In this book I wish to concentrate on some very basic shifts in the amount of physical and organic matter humans use. Humans transform matter. We make artifacts from stone and wood and clay and metal. We domesticate plants and animals, light fires and burn fuels, and make machines. The amount of this human-made stuff in use at any given time has increased.

For at least seventy thousand years, anatomically modern humans—people biologically like us in every way—lived in small, mobile groups of ten to thirty people, aggregating from time to time, sometimes producing wonderful wall paintings and magnificent implements. Their success and mobility were possible partly because they carried very little with them. They wore clothes made of skin tied together with sinews and plant cords. They had baskets and skin containers, and over time they added bone tools such as needles. They had wooden spears and bows, as well as tools and weapons made of chipped stone, such as flint and obsidian. They lived in cave entrances or in huts made of plant matters or bones from wild animals. You could place on a small table all the material belongings of a man or woman who lived thirty thousand years ago. They had very little stuff. Moreover, when the stuff ran out, wore out, or broke, it was easy to replace. Most of the materials used were organic and easily found and remade.

About ten thousand years ago, the amount of stuff in peoples' lives increased dramatically. As Colin Renfrew put it, "Human culture became more substantive, more material."[5] People following a mobile existence could accumulate only so much, but once they settled in one place, the potential for surrounding themselves with material things increased. Or we might turn this around: Perhaps increasing material accumulation forced people to settle down and start farming. A striking amount of new stuff became part of their lives. Between 12,000 and 7000 BCE in the Middle East, people started living in permanent houses made of sun-dried mud brick[6] with enclosed living and storage areas as well as burial and ritual spaces. By 8500 BCE some houses had two stories and substantial roofs made of clay and reeds and timbers. In the houses were stored cereals, now domesticated and changed by human intervention, as were the flocks of domestic sheep, pig, and cattle. The animals provided humans with large amounts of meat that could be owned, stored, dried, and used in feasting. Ground stone implements were ubiquitous by 12,000 BCE and were used to make a variety of querns and pounders and abraders; finer stones were ground into polished axes used to cut down trees to make timbers for houses and burial chambers.

Pottery made of fired clay was invented, providing storage, cooking, and eating containers for sedentary communities; fired clay was also used for pot stands, figurines, and stamp seals. Weaving implements in the form of spindle whorls appeared, suggesting a range of cloth goods that rarely survive, made from wool and flax. There was an increased variety of tools (including spoons and forks), dress fittings and ornaments made of animal bone, as well as beads and necklaces made of bone, shell, and stone. People expanded the range of wooden containers to include bowls and cups, and used an increasing diversity of baskets. It would not be possible to place on a small table all the belongings of a household from this period.

In Europe there was another peak in the amount of stuff made and used by humans in the Roman period (by which time people actually had tables!), but the most marked increase has occurred since the Industrial Revolution. And this increase has spread out across the globe to many, but not equally to all, places. Many of us could fit only a tiny fraction of our possessions onto a small living-room table, and a minute fraction of all the material resources mobilized to produce and maintain the consumer goods on which houses, cities, nation-states, and global communications rely. Today the largest single machine in the world is the Hadron Collider, a twenty-seven-kilometer ring of superconducting magnets used to study subatomic particles. Constructing this machine required the collaboration of over ten thousand scientists and engineers from more than one hundred countries, and hundreds of universities and laboratories. Its information network connects 170 computing centers in thirty-six countries. The electricity to run the machine costs $23.4 million annually. This is "energy capture" on a massive scale. We have come a long way from the first stone tools.

Another way of exploring this trend is to contrast early human-made things with those produced today. This gives us only a crude indication of change in the amount of material culture, but the contrasts are nevertheless instructive. Take, for example, the earliest sickles used to cut grasses.[7] Starting from around 12,000 BCE in the pre-Neolithic Natufian culture, these simple flint (or occasionally obsidian) blades were set in hafts of wood, antler, or horn and glued with bitumen (figure 1.3). Although obsidian was often obtained over great distances, flint could often be found locally. Osamu Maeda argues that sickles were originally developed as a cutting tool for raw materials such as reeds and sedges.[8] Later, over the course of the ninth millennium BCE in the Middle East, sickles were transferred to agricultural harvesting. These early sickles were easily mended or replaced, or done without: People could also harvest grain by uprooting the stalks or beating the seeds into baskets. Over time, people became dependent on more efficient harvesting tools. Metal sickles were used in the Bronze and Iron Ages in Europe, and scythes had appeared by the Late Bronze Age. There is some evidence of reapers pushed by horses in the Roman world, and the world's first patent for a reaping machine was issued in England in 1799.[9] The modern agricultural industry has become dependent on massive combine harvesters, so named because they "combine" reaping, threshing, and winnowing into one process. The labor savings on the farm are considerable, but these machines often cost four hundred thousand dollars and have over seventeen thousand parts that are made all over the world—and are distributed to customers by, for example, the John Deere company through its global parts distribution network strategy. The latter company deals with eight hundred thousand parts for its various machines.

Or take the example of spinning (figure 1.4). Domesticated

Figure 1.3 Neolithic stone sickle blades and modern combine harvesters.
Source: Wikimedia commons.

sheep with woolly coats appeared late in the Neolithic period in the Middle East, and the first spindle whorls appeared around 7500–7000 BCE.[10] These small, innocent whorls, used for spinning wool into yarn, were easy to make from clay or stone, and the wooden sticks and the sheep's wool were available locally and easily replaceable. We can follow the development of spinning technology from the medieval spinning wheel to the various machines used to spin wool, linen, and cotton. In the case of cotton, the first spinning machine with multiple spindles was the spinning jenny, invented in England in the mid-eighteenth century; then, in the 1780s in Manchester, spinning machines called water frames were powered by waterwheels, leading to spinning mules powered by water or steam, leading directly to the vast spinning machines in textile factories today. Nowadays making, producing, and selling cotton T-shirts is a massive global enterprise that, on the one hand, employs millions of people and connects the world, but also has serious environmental impact.

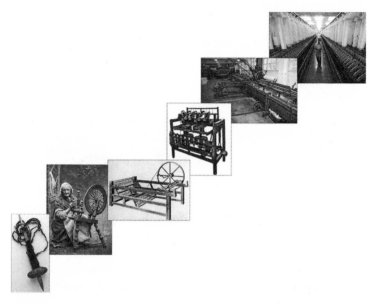

Figure 1.4. The development of spinning technologies from Neolithic spindle whorls to spinning wheels, spinning jennies, water frames, mules, and modern factories. Source: Getty Images.

Another classic example is the wheel (figure 1.5). Although textbooks often say it was invented in Europe and Asia in the fourth millennium BCE, its origin is in fact very difficult to pinpoint. The idea of a rotating axle and wheel has many sources, including spindle whorls that, as we have already seen, began in the eighth millennium BCE. Even earlier, in the Upper Paleolithic, the rotating bow drill was used to make holes in beads, and earlier still, sticks rotated by hand were used to make fire. The potter's wheel emerges at about the same time as wheels used in transport, but it is very difficult to say that the two devices were connected. What is clear, however, is that after this early start from multiple origins, humans have increasingly made use of the multiple affordances of wheels—for transport, weaponry, energy production, clocks and watches, tools and lathes, spinning wheels and machines, musical instruments, and a thousand other things. Envisaging present-day society without the wheel is impossible. For example, we have become thoroughly dependent on the car, have become tied in to a global trade in parts, and have had to deal with the effects on global climate of the massive production of greenhouse gases. Figure 1.5 is an effort to summarize the long-term global proliferation of uses of the wheel. As with harvesting and spinning technologies, early simple things have become greatly elaborated and have increased in number, embroiling humans ever more deeply in complex relationships with each other and the global environment.

It is difficult to draw out all the proliferations because the mass of connections becomes so large. But for an archaeologist it is possible to sketch the links on one site. For example, at the Neolithic site of Çatalhöyük that I have been excavating in Turkey, the earliest pottery was made from local clays using simple technologies, and the pots were used as containers. Later the pottery was made with greater skill and used for cooking. Still later they came to be used for storage, as well as consumption and social display, as seen in the painted decoration. Through time the affordances of pottery are gradually realized and the connections increase (figure 1.6). More and more things became tied to or dependent on pottery, just as more and more things got tied to or dependent on

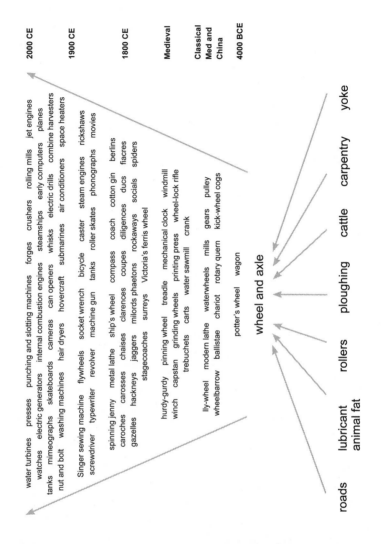

Figure 1.5. Some of the context that led to the adoption of the wheel and axle in Eurasia in the fourth millennium BCE, and the gradual proliferation through time of the uses of the wheel and axle.
Source: Author

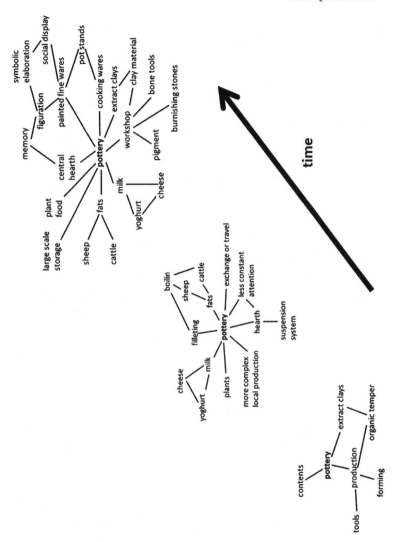

Figure 1.6. Gradual expansion of the uses of pottery in the early, middle, and late phases of the Neolithic site of Çatalhöyük in Turkey in the seventh and sixth millennia BCE. Source: Author.

wheels, spinning machines, and harvesting tools. This set of dependencies is what I define as an "entanglement."

In this book I describe human evolution in terms of increases in the amount of humanly modified materials per head of population, increases in the rate of change of this measure, and increases in human-thing entanglement. I mentioned above that you could place on a small table all the material belongings of a man or woman of thirty thousand years ago. The average US household today has three hundred thousand things, from paper clips to ironing boards,[11] and most US homes have more TV sets than people. Frank Trentmann in his book *The Empire of Things* has charted the rise in the amount of consumer stuff in Europe and China from the 1500s onward.[12] But as with energy capture, there is a great deal of variation. In my work among contemporary pastoralist and small-scale agricultural societies in Kenya, Zambia, and Sudan in the 1970s and 1980s, I counted the number of artifacts[13] and found between ten and fifty items per household. The 12 percent of the world's population that lives in North America and Western Europe account for 60 percent of private consumption spending, while the 33 percent living in South Asia and sub-Saharan Africa account for only 3.2 percent.[14] About 2.8 billion people on the planet struggle to survive on less than $2 a day, whereas in the United States the median household income was $57,617 in 2016.[15] Any account of the increase over time in the amount of human-made things per person must also consider the divergence between those with a lot and those with very little.

Why do these evolutionary trends happen? In this book I start by considering how this question is answered by theories of progress, by theories of biological evolution, and briefly by complexity theories. I then outline an entanglement theory that I believe offers a more adequate answer. But I need also to explain why I focus on increases in material stuff and human-thing dependence rather than on other evolutionary trends. Part of the answer is that other evolutionary trends—such as increases in specialization, differentiation, complexity, inequality, heterogeneity, and integration—are all linked to the increase in material stuff. It is even difficult to discuss human genetic change over the last 1.5 million years without referring to the material niches humans have constructed for themselves over that period.[16]

A more complete answer, however, is that today humans as a species face intractable problems that are a direct result of our increasing energy use and our heavy dependence on material technologies. The examples in this chapter were chosen with these problems in mind. The stone sickle has grown into an agricultural economy that, according to the Food and Agriculture Organization of the United Nations (FAO), is responsible for 18 percent of the total release of greenhouse gases worldwide. The spindle whorl has led to a cotton industry in which the production of one T-shirt uses seven hundred gallons of water, the amount of carbon dioxide emitted to produce one pair of jeans is equivalent to driving a car seventy-eight miles, and cotton crops receive 25 percent of the world's pesticides and herbicides. As for the wheel, the US transportation sector, which includes cars, trucks, planes, trains, ships, and freight, produces nearly 30 percent of all US global warming emissions.[17] The human dependence on and attachment to things we have created are directly related to the problem of global warming.

We tend to deal with such problems by inventing yet more things. Genetically modified crops are sold with the promise that they are good for the environment. Petroleum-based synthetic fibers have started to replace cotton manufacturing. Large companies have invested in fuel-efficient vehicles. More fanciful responses to global warming also focus on technological innovation. One solution is for factories to pump sulfur dioxide through a fire hose up to zeppelins hovering sixty-five thousand feet in the air. This aerosol would shield the planet from the sun, keeping it cool. But all such geo-engineering schemes have negative consequences that would require yet further intervention.[18]

This is the same thing we have always done. In this book I argue that, as a species, we have dealt with many problems by seeking technological solutions that then cause further problems, and so on in an endless spiral. Our dependence on material things has drawn us into ever more entanglements and inequalities. At present the problems of climate change seem intractable. There is much pushback at global climate agreements. As a species we seem to think that the most viable solution is technological. I return at the end of the book to consider whether there are alternatives. For the moment we need to try and understand what has led us to this ever greater dependence on things.

The Idea of Progress

ONE OBVIOUS ANSWER to the question of why humans have moved in the direction of a greater use of material things is simply that they wanted to. Of course, this quickly leads to the next question: Why did they want to? We can sidestep this second question by asserting that there is just a basic human drive to better oneself, and this desire is intimately tied to material technologies. One distinctive aspect of humans is that, compared to other animals, they make more complex tools with which to capture energy from the environment, and they keep improving those tools. More and better things result in better housing, more reliable food, more medical treatment, travel, luxury, and education. Accumulation of things is the accumulation of wealth, and it allows the exercise of power. The accumulation of power supports specialists and artists who expand the range and depth of human experience, both material and spiritual. So is it not obvious that humans want to progress and better themselves through the use of things?

In fact, many people argue that the idea of progress is a recent invention, a product of the rise of capitalism in the eighteenth and nineteenth centuries CE in Western Europe and North America.

Certainly nineteenth-century Britain was drenched with the ideas of evolutionary progress, "improvement," or "betterment." But a good case can be made that the idea of progress has great antiquity and was present in the Greek and Roman worlds.[1] The early Christians, especially St. Augustine, described a necessary movement toward spiritual perfection—a millenarian movement of worldly struggle leading toward a promised golden age. Similar ideas of a necessary unfolding according to God's plan are found throughout medieval Europe—as well as in recent times. In the 1920s, archaeologist William Foxwell Albright described the role of Israel in the history of civilizations in terms of a directionality grasped by a consciousness beyond our ability to comprehend the indecipherable will of God in history.[2]

The opening of the Americas and other parts of the world to Europe in the fifteenth to seventeenth centuries CE challenged the West to make sense of the differences between these newly found societies and its own. How could one locate them socially, economically, and spiritually? The most common way was through the idea of progress—that "they" were now as the West had been. Seeing "them" as prehistoric savages reinforced European notions of superiority and Manifest Destiny and justified exploitation and enslavement.

The idea of progress reached its zenith in the West in the 1750–1900 CE period. Progress underpinned colonialism and imperialism and other key ideas of this period, such as freedom, equality, social justice, and popular sovereignty.[3] In a context of progress, these ideas became not just desirable but historically necessary and inevitable. Comte, Marx, Hegel, and Spencer could all write histories of a slow, gradual, continuous, and necessary ascent to some end. The idea of progress became secularized—

wrested from a divine plan and seen as a natural process subject to scientific analysis.

Two trends in the development of ideas of progress had clear political ends. The first is the celebration of the ideas of freedom and individual liberty. According to Adam Smith in *The Wealth of Nations* (1776), wealth and prosperity derive from allowing every person to better his or her own condition. The famous "invisible hand" that Smith thought drove societies forward was the pursuit of private interest leading to a maximum of public good. The drafters of the 1776 US Declaration of Independence were also steeped in the idea of progress. They were strongly influenced by Protestant millennialism and the idea that one nation, the United States, could be a redeemer in the overall progress of mankind. Adams, Franklin, Jefferson, and Paine all believed in the gradual betterment of the arts, sciences, technology, and industry, unfettered from the yoke of heavy-handed government. In Europe there were many, such as Immanuel Kant and John Stuart Mill, who argued for "laws of progress" that would gradually bring welfare and freedom through the promotion of individual liberty. But it was Herbert Spencer who was the most influential proponent of a link between individual liberty and progress. He saw a movement from homogeneity to heterogeneity as monolithic, repressive, and static forms of social organization were replaced by diversified, individualistic forms. Legislation by the state, he argued, cannot solve individual human problems or plan for society. He introduced the phrase "survival of the fittest."

The second trend in eighteenth- and nineteenth-century ideas of progress was a more critical and utopian idea of the role of power and the state in producing progress. For Rousseau, progress meant movement toward a society in which all inequalities were erased in favor of the common good. Inequality had started with agriculture and metallurgy, which led to property and servitude. Private property created social interdependencies from which humans could not liberate themselves, and the state arose from the resulting conflicts of class, nation, and territory. All this was a diversion from true human progress toward equality. Auguste Comte saw individualism as "the disease of the western world," and he favored a spiritual authority that would direct public opinion and establish the principles on which stable social relationships would be based. Marx proposed a broad evolutionary theory in which conflict and contradiction would produce class struggle and lead inexorably toward the end of capitalism. He and Engels stressed that human history progressed through savagery, barbarism, slavery, feudalism, and capitalism to its ultimate utopian state. In their view, the intermediate steps in the march toward communism required the imposition of power: Marx talked of the dictatorship of the proletariat.

We thus see, in the heyday of Western ideas about progress, a struggle over the political use of the term. On the one hand, notions of individual liberty made free markets and mercantile capitalism inevitable, and both underpinned the formation of the United States and justified Western imperial conquest of "primitives." But progress also led beyond the present world to a utopian ideal of equality or spiritual fulfillment guaranteed by the power of the state. The idea of progress thus underpinned a wide variety of social and political movements and laid the foundations for the modern world.

In the twentieth century, Henri Bergson offered a more sophisticated idea of a forward impetus. He criticized the "false evolutionism of Spencer" and rejected both adaptationist and goal-directed theories.[4] He recognized that many evolutionary pathways can be taken, and that some are contrary, deviations, arrests, and setbacks. "We must recognize," he wrote, "that all is not coherent in nature." Change was motivated by something more basic and general than the causes proffered by the likes of Spencer, Comte, and Marx: "an *original impetus* of life" that pushes the generations forward. Life is just the creative tendency to act on inert matter, but the direction of this action is not predetermined. For Bergson, the "original impetus" was "an internal push that has carried life, by more and more complex forms, to higher and higher destinies." Evolution was "a creation unceasingly renewed." The living tendency to create, develop, and grow moved down divergent pathways so that differentiation and specialization could occur. And for Bergson, despite his insistence on indeterminacy, there was no doubt that the life-producing impetus produced progress in the sense of a continual advance.

Bergson saw the creative impulse pushing forward by taking energy from matter and putting it to use. "The impetus of life, of which we are speaking, consists in a need of creation. It cannot create absolutely, because it is confronted with matter, that is to say with the movement that is the inverse of its own." A later writer, Leslie White, who also saw differences between biological evolution and the physical processes of matter, had a significant impact in archaeology and anthropology. You can find strong resonances of his ideas in the work of Ian Morris, which we discussed in chapter 1. Like Morris, White saw social evolutionary progress in terms of energy capture and described high and low cultures in terms of the amount of energy they can harness.[5] Higher forms, the greater civilizations, use energy to expand and become more complex, with greater specialization and integration. Culture advances with increases in the per-capita amount of energy harnessed.

White saw a basic contradiction between the physics of matter, which follow the first and second laws of thermodynamics (the conservation of energy and the gradual entropic dissipation of energy), and biological processes, which go in an opposite direction. He argued that according to the second law, "the universe is breaking down structurally and running down dynamically; i.e., it is moving in the direction of lesser degrees of order and toward a more uniform distribution of energy. The logical conclusion of this trend is a uniform, random state, or chaos," which was the state of maximum entropy. The universe moves toward a uniform state of disorder. White then writes very dramatically that "in a tiny sector of the cosmos, however, we find a movement in the opposite direction. In the evolution of living material systems, matter becomes more highly organized and energy is raised from lower to higher levels of concentration." Living beings manage this by sucking order and energy from the environment through chemical processes such as photosynthesis. "Living systems are means of arresting, and even of reversing, the cosmic drift toward maximum entropy." They are energy-capturing systems.

White argued that the life process tended to augment itself. It did this by the multiplication of numbers through reproduction and by developing higher forms of life. By "higher forms" he meant that animals are more highly developed thermodynamic systems than plants, and mammals more than reptiles. In human social systems as well, there are higher and lower levels of energy capture. Cultural systems, like biological systems, "extend themselves"—quantitatively as when people multiply or tribes divide, and qualitatively by developing higher levels of organization and greater concentrations of energy. Why this augmentation happens is not clear in White's writing. At times he seems to assume—rather like Bergson—an impetus that moves the life process forward. Thus, "any species will tend to make full use of any means that it possesses to make life secure and also to expand and extend itself." At other times, augmentation seems to occur because of the competition to survive. The advantage in the evolutionary struggle for survival goes to those organisms that are most efficient at energy capture. Certainly, when he discusses evolutionary developments such as farming and urbanism, White gives much causative play to pressure on resources, population pressure, and climate change. Many evolutionists have argued that in the struggle for survival it is the best adapted and the most able to harness energy that wins. I return to these types of competitive adaptationist arguments in the next chapter. For the moment, it remains unclear, given the increased energy these societal shifts require, why competitive struggle does not also favor groups with less energy capture.

The social evolutionary theories of progress put forward by White and others became unpopular in the latter part of the twentieth century. For example, Bruce Trigger criticized the way these lawlike evolutionary theories, though perhaps purged of nineteenth-century triumphalism, nevertheless created a narrative that ignored the specific histories and agencies of marginalized groups such as Native Americans.[6] General progressivist ideas are still found today in evolutionary theory. For example, Bruce Smith has argued that niche construction theory can explain "initial domestication not as an adaptive response to an adverse environmental shift or to human population growth or packing but rather as the result of deliberate human enhancement of resource-rich environments."[7] The notion that humans consciously enhanced the density and productivity of desired resources is repeated by others as well—but it is precisely this "deliberate human enhancement" that needs to be explained rather than assumed.

Another example of the way a progressive drive appears to underlie some contemporary evolutionary theories is the assumption that populations increase. For example, we saw that White assumes that people just multiply where and when they can, and today's niche construction theory often depends on population increase as a macroevolutionary process that generates change. In fact, it has proved very difficult to use population increase to explain long-term change.[8] It turns out, for example, that population and resource pressures were not primary causes of the adoption of agriculture in the Middle East, and in any case we would still have to explain why populations increase. Human populations have often declined, and today, rates of population growth vary dramatically around the world. It is difficult to argue that "other things being equal," populations will always increase, since other things rarely are equal. If populations rise or fall, it is because of specific circumstances such as resource pressure, the need for labor, social and political conditions, religious belief, and so on. We cannot assume that populations always "just do" increase.

As another example of the ways in which contemporary scholars promote directionality, John Stewart argues that human groups show an overall progress toward cooperation and management—that is, toward hierarchical systems that support cooperation and suppress cheats.[9] He starts by suggesting that there are benefits to groups if individual entities cooperate. Hierarchical systems of management and control are an important mechanism for suppressing cheats. But the tendency to cheat will remain, and there may be instability between different hierarchical systems. Higher levels of organization and management are thus needed to deal with between-group instability and free-riding—and so on in an endless spiral of cooperation on larger and larger scales. This scenario is intuitively attractive because it seems to explain the growth of human cooperative systems from small bands to the United Nations. There is an apparent directionality in the degree to which humans are willing to submit to chiefs, royalty, states, and governments.

Yet cooperation can be defined in many ways. Small hunter-gatherer societies often have strong leveling mechanisms and emphasize the sharing of resources. On the other hand, the dysfunction of many national and international bodies does not suggest successful cooperation to any great degree. And cooperation produced by higher-level management usually involves coercion; are we to say that the master and slave "cooperate"? The scale, type, and degree of cooperation seem to vary over time and place. It is hard to see an overall trend.

INCREASING COMPLEXITY

White's account of evolutionary change can be seen as a narrative of increasing social complexity. The notion that there are more levels of organization as societies move from bands to tribes to chiefdoms and states is one example of an idea that was foundational in many social disciplines into the mid-twentieth century. For example, Julian Huxley saw increasing complexity in biological evolution in that a bird or a mammal is more complex than a fish, a fish more complex than a worm, a worm more complex than a polyp, and so on. "In the human sector, a new complexity is superimposed on the old, in the shape of man's tools and machines and social organization. And this, too, increases with time. The elaboration of a modern state, or of a machine-tool factory in it, is almost infinitely greater than that of a primitive tribe or the wooden and stone implements available to its inhabitants."[10] Huxley also noted that the rate of evolution toward greater complexity is forever accelerating; I return to this issue later.

Discussion of complexity theories certainly has a place in understanding the current state of world biosocial systems.[11] One difficulty with such theories is how "complexity" is to be defined. Is it best measured as numbers of different types or species? If so, the difficulty is compounded, at least in archaeology, since there are many ways to group artifacts into types; some archaeologists are "lumpers," producing large general groups, and others are "splitters," producing ever more refined and detailed classifications. Is it the diversity of types? Or the number of levels of organization? Is complexity related to specialization, degrees of specialization, and levels of integration? Given these problems of definition, it is difficult to respond to the general idea that complexity tends to increase, or to say why.

Can modern complexity theory—meaning the science of complex systems, whether physical, biological, or social—help explain overall directionality? Complex systems are often defined as self-organizing. A self-organizing system can certainly make itself more complex, but since this will happen only under certain conditions,[12] it is difficult to argue that self-organizing complex systems always make themselves more complex. Nonlinear interactions among variables often produce sudden and unanticipated effects, and one could argue that the accumulation of such effects will eventually lead to greater complexity. Daniel McShea and Robert Brandon have suggested a "zero-force evolutionary law" that states that "diversity and complexity arise by the simple accumulation of accidents."[13] For example, if you start with a clean white picket fence, it will gradually get dirty, and paint will start to peel off, thus creating diversity and complexity in picket fences. Some archaeologists have argued that this type of "random drift" can lead to change in pottery styles.[14] But changes in pottery styles and picket fences are not entirely random; the accidents have to be dealt with, adopted, or selected for. For example, some house owners carefully maintain their picket fences, and if a tree falls and breaks the fence, the owners may quickly fix and restore it. Whether complexity arises out of nonlinear or accidental events thus depends on many other factors. It remains difficult to see why systems should necessarily become more complex through time. A recent volume of papers by physicists, biologists, and other scientists studying self-organizing systems found no consensus on either the questions "What is complexity and how do we measure it?" or "Does complexity increase?"[15]

We could perhaps define increases in the amount of human-made things, in their diversity of form and function, and in their degree of interdependence as increases in complexity. Over time, as we saw in chapter 1, tools and machines do get more complex. Yet while complexity theories may help us define and describe these increases, there seems to be no accepted underlying theory that would explain why complexity should increase as a general rule. As we will see, entanglement theory lets us turn the question around and ask how human-thing dependencies might pull complexity in a general direction.

DIRECTION ISN'T "PROGRESS"

The idea of progress seems to embrace four themes. First, there are stages of development. Second, societies move through the stages gradually and cumulatively. Third, the later stages are superior to the earlier ones, and fourth, the stages are natural, inexorable, or necessary. What does "superior" mean? There are many ways of defining progress toward some superior state. (We shall see the same difficulty in biology in chapter 3.) "Progress" often means an increase in individual liberty or an increase in technology and knowledge, especially scientific objective knowledge. "Superior" sometimes means greater moral or spiritual happiness, well-being, serenity—a greater perfection of human nature. But many have commented, sometimes angrily, that technological advancement is linked to moral decline. We recognize the advance of medicine but also see the need to manage it—hence the burgeoning of medical ethics. In recent centuries progress in the Western world was positively linked to science, democracy, and liberalism, but it also gave rise to totalitarianism, supremacist racisms, colonialism, and imperialism—with archaeology providing a scientific basis for regarding other societies as primitive or lost in time.

These aspects of the idea of progress were all mired in special interests, whether the supremacy of Athens, the centrality of the church, or the imperial dominance of the West. The notion that stages of development followed from each other gradually and inexorably served the interests of those who benefited from established power and privilege, and who feared that revolutionary change would undermine it. The superiority of later stages justified the policies of imperial expansion, slavery, and genocide, and the consensus that these stages were inexorable placed those policies beyond reproach. The notion that the present is the apex of progressive development is the ideology of those in power.

But the main difficulty with ideas of progress is that they are teleological. The thing to be explained (societies have attained objective science, complexity, democracy, or freedom) is also the cause (societies progressed in order to achieve objective science, complexity, democracy, or freedom). Certainly there have been societies that pursued specific goals, such as sharing resources with other living beings in the environment, amassing more cattle, creating equality, pursuing the will of God, or creating wealth, freedom, and democracy, and these goals contribute to short-term change. But it is inadequate to say that humans overall have progressed because they wanted to. This would assume that the desire for progress is somehow inherent in the human species. Yet there have been long periods, such as most of the Paleolithic, in which change seems to have been very slow indeed. And parts of the world did not show marked material accumulation through the Holocene. Whether societies develop a strong notion of progress seems to depend on context; we cannot assume a universal inherent drive toward self-betterment.

Given these problems with ideas of progressive development, how can I justify my focus on increases in stuff? In chapter 1, I said I would discuss increases in the amount of humanly modified materials per head of population, increases in the rate of change of this measure, and increases in human-thing entanglement. Doesn't this perspective derive from a focus on accumulating consumer goods, on material wealth, on technology as a source of progress and development? There are other perspectives, often religious or philosophical, that eschew such values and focus on immateriality, meditation, and transcendence. Am I not in this book just providing an origin story that explains the rise of contemporary consumerism and capitalism?[16] Undoubtedly the answer to this is "yes." But I hope it will also become clear that my purpose in exploring this rise in materialism is critical. I argue that as much as things and technologies have played positive roles in human development, they have also drawn us into unacceptable inequalities and unsustainable adaptations. It is unclear on the overall balance sheet that more things actually represent progress.

Nor do I propose that humans have an inherent drive to accumulate more stuff. Rather, I argue that stuff draws humans into an engagement that leads to directional change. This is not "progress." The only inherent assumption I make is that humans depend on things. This is no more than saying that *Homo sapiens* is also *Homo faber*—a tool maker. Dependence on material things is inherent to the human species and has drawn that species in a particular direction. If at particular historical moments, humans have a drive to accumulate more stuff, this is because of the direction humans have been drawn into by human-thing interactions—what I later describe as "entanglements."

Does Biological Evolution Provide an Answer?

D OES THE PARALLEL between cultural and biological evolution help to explain the directionality I have described? Is the increase in material stuff simply an extension of an overall increase in biological complexity? Many have compared the transmission of cultural traits to the inheritance of genes.[1] There are, of course, doubts about whether cultural and biological evolution are really comparable.[2] One problem with the analogy is that genetic change happens much more slowly than cultural change, which can happen within generations. Another problem is that while biological organisms reproduce, cultural objects have to be reproduced by humans; they cannot reproduce themselves. Nowadays we can write computer code and generate artificially intelligent machines that self-reproduce, but only within constraints we set. The replication of cultural objects is not equivalent to reproduction of organisms.

And yet we might still argue that increases over time in the amount of material stuff made by humans have made our species more complex. The growth of material culture is thus an example of the greater complexity of species through time. Surely the human case proves that biological evolution has an overall directionality? Arguments derived from Darwinian theories are attractive because they avoid the goal-directed assumptions that underlay the theories of progress discussed in chapter 2. Within a Darwinian frame, evolution has no goal. Biological traits are selected that enhance fitness for the particular environment in which a population lives.

The specific directionality that occurs as organisms adapt to local environments is widely accepted in biology. Directional selection occurs when individuals with traits on one side of the mean in their population survive better or reproduce more than those with traits on the other side of the mean. The increase in prevalence of advantageous traits gradually yields organisms with maximum fitness in a particular environment. Darwin gave examples: Faster wolves are more successful at hunting deer, and flowers that produce more nectar are more successful at attracting pollinating insects. Because local environments vary, this process can lead to divergence and the formation of new species. We can, for instance, imagine an environment in which speed is not the most important trait for a wolf. If herds of large, slow herbivores entered the area, the wolves with heavier, more muscular bodies or with a better ability to hunt cooperatively might gain a survival advantage over the leaner, faster ones. Over many generations, the local population of wolves might become on average slower, heavier, and more social.

But this process of local adaptation does not imply any general direction. Specific directional selection does not necessarily lead toward increasing complexity, greater intelligence, an overall process of accumulation, or increasing differentiation/integration. At any moment, the local environment could begin to favor simpler, less differentiated forms.

GENERAL DIRECTIONALITY IN BIOLOGICAL EVOLUTION

It is often argued that in Darwin's theory of natural selection, neither progress nor cumulative development plays a significant role. But Roger Nisbet showed that Darwin did have an overall notion of progress toward more perfect forms: He believed the continued action of natural selection leads to progressive development.[3] Stephen Jay Gould also pondered those segments of Darwin's writings in which Darwin proposed an overall progressive development.[4] Gould explained these by supposing that Darwin unconsciously adopted the ideas of progress prevalent in Victorian imperial Britain. But Gould himself believed very strongly that genetic "variation itself supplies no directional component." In a famous dispute with the paleontologist Simon Conway Morris, he argued that if we were to rewind the tape of evolution from the beginning, randomness and contingency would lead to completely different outcomes.

One example of general directionality that might be claimed is "Cope's Rule," which states that there is selection for increasing body size. The first animals to evolve were tiny, whereas many modern animal species are large. Was this increase in size due to active selection, or to some more random process? Noel Heim and his research team explored this hypothesis in marine animals and found that body volumes have increased by more than five orders of magnitude since the first animals evolved.[5] In addition, their modeling suggests that such a massive increase could not have emerged randomly. There remains much debate about this rule, and it does not hold true at all taxonomic levels or in all environments. For instance, when an animal population is isolated on an island, its body size tends to shrink.

One of the problems with identifying general progress in evolution—just as with cultural evolution—is that a lot depends on how you define progress. In biology, it can be measured in terms of relative information content of DNA, ecological dominance, invasion of new habitats, expansion of life, replacements, increased specialization, increased complexity (itself difficult to define—see below), or increase in general energy, to name just a few. "By certain criteria," writes Michael Benton, "flowering plants are more progressive than many animals."[6] Joseph Fracchia and Richard Lewontin have written that "no agreement can be reached on how to measure complexity independent of the explanatory work it is supposed to do."[7] It thus remains difficult to identify general trends.

Yet many people assume some notion of overall evolutionary progress, generally with *Homo sapiens* at the top of the pile. One reason given for this progress is that competition leads to improvement. For example, it is often claimed that biological complexity has increased over time[8] because more complex organisms are able to survive better (although a claim might be made for the adaptive advantages of simplicity too). But after reviewing a large amount of evidence for interspecific competition and replacement, Michael Benton argued that the history of life does not show an improvement in competitive abilities over time.[9] There is no good evidence that the appearance of key adaptations leads to large-scale competitive replacements of one species by another. "It has not been demonstrated," Benton wrote, "that increases in morphological complexity correspond to better adaptation, that life has expanded in anything but an opportunistic way, that evolutionary trends show improving competitive ability, or that adaptations become increasingly effective through time."

EVOLUTION AS A CUMULATIVE
BIO-SOCIO-MATERIAL PROCESS

But perhaps evolution demonstrates progress in a rather different way: New developments build on old. In other words, evolution is cumulative. While there is no teleological drive toward improvement and progress, perhaps there is directionality caused by the fact that natural selection has to work on what is already there. Henri Bergson saw biological evolution's cumulative nature as a kind of directionality.[10] Genotypes tend to build on themselves instead of starting over. The neo-Darwinian synthesis proposes that the only significant constraints on evolution are imposed by the environment, but recent studies demonstrate that genotypes also impose constraints, and many of these are legacies of evolutionary paths taken long ago. As Stephen Jay Gould rather colorfully put it, elephants are unlikely to produce genetic variations that lead to wings. All species have structural forms that limit the possible range of variation.[11] There is a bias in the expression of phenotypic variability due to the interactions that characterize developmental systems.

Biological evolution is also cumulative in another sense. Humans, primates, birds, and many other animals pass on information from generation to generation. They accumulate cultural information over time. In recent decades, many evolutionists have argued that there are two different forms of transmission: the biological inheritance of genes and the cultural replication of information. Examples of the latter include blue tits learning from other blue tits how to get through the metal caps of milk bottles, and Japanese monkeys passing on the idea of washing sweet potatoes.[12]

Most biologists and social scientists have tended to see natural selection and cultural transmission as separate if analogous and parallel processes. Richard Dawkins argued that the transmission of genes was absolutely separate from the transmission of cultural memes. Robert Boyd and Peter Richerson took the same approach in their dual inheritance theory. Bill Durham has argued for four transmission channels: cultural, biological, ecological, and social. Although he saw complex interactions among them, he argued that it was methodologically necessary to separate them.[13]

The main difficulty with dual or multiple inheritance theories is that they tend to view culture as "ideas in people's heads." More recent scholarship has shown that culture is always to some degree constructed and history is always to some degree invented. Culture cannot be seen simply as a set of traits passed down from one generation to the next,[14] nor is it possible to separate off a sphere of culture that is transmitted. Rather, culture is embedded in daily practices and practical logics. For example, cultural rules about how to build houses and arrange the activities inside them are tied to the physical and chemical properties of wood and brick and stone, as well as to biological agents such as bacteria, rodents, and flies that inhabit the house space and enter into human decisions about cultural rules and practices.

Richard Lewontin's "triple helix" of gene, organism, and environment provides a perspective on biological evolution that is more in tune with current understanding of culture and history. Lewontin emphasizes that an "organism is not specified by its genes," that an independent environment cannot be defined, and that notions of adaptation are inadequate in that they seem to place the organism in opposition to environment.[15] He construes evolution as construction, much in the way that modern theorists describe social practice, argues for the importance of particular histories resulting from small contingent differences, and shows that mutation is not random. Gene, organism, and environment offer a fluid set of relations in which it is difficult to separate out individual causes or effects. Earlier, Bergson had taken a very similar approach to adaptation when he criticized the idea that there are preexisting niches into which organisms fit. "In the adaptation of an organism to the circumstances it has to live in," Bergson wrote, "where is the preexisting form awaiting its matter? The circumstances are not a mold into which life is inserted and whose form life adopts: This is indeed to be fooled by a metaphor. There is no form yet, and the life must create a form for itself, suited to the circumstances which are made for it."[16]

Eva Jablonka and Marion Lamb show how gene-centered views became undermined once the complexity of the relationship between genes and characters became clear: No gene could be found for obesity, sexual preference, criminality, religiosity, and so on.[17] The relationships among DNA, RNA, and proteins are not simple linear pathways. There are regulators that determine how genes are turned on or expressed; these regulators are influenced by the environment, and they are heritable. Gene expression is controlled or managed or responsive to context; there is evidence of plasticity, atavism, cryptic genes, and genetic accommodation. Within the complex assemblages that make up an organism, surprising interactions can occur. For example, aphid body color and other phenotypes are transmissible through the alleles of their symbiotic bacteria.[18]

So genes (and by analogy memes) do not replicate themselves; they need the cell, the organism, the environment, the heterogeneous community of inter- or intra-actants.[19] There is no single agent and no single cause. There is the onward-goingness of life, but agency is dispersed through the genome and into broader material, social, and cultural entanglements. The genome comes to be seen as responsive—what Evelyn Fox Keller calls the reactive genome.[20] She argues that it is no longer adequate to describe a simple causal chain from genotype to phenotype. The genome is able to respond to outside stimuli through the mechanisms of epigenetic inheritance such as chromatin remodeling and methylation. Noncoding RNA is involved in many forms of genetic regulation. Unlike DNA, RNA sequences are malleable and can be rewritten and reinscribed, and through reverse transcription, these changes can be incorporated into DNA. So noncoding RNA not only allows gene expression to respond to variations in the environment but allows the environment to influence the genome itself.

Thus it makes no sense, Keller writes, to separate biological from cultural factors. "We have long understood that organisms interact with their environments, that interactions between genetics and environment, between biology and culture, are crucial to making us what we are. What research in genomics shows is that, at every level, biology itself is constituted by those interactions— even at the level of genetics." It thus makes no sense to separate different transmission channels, different pathways. Culture, material culture, and the social are not separate things running parallel to the biological and the genetic. Rather, the cultural, the social, and the material enter directly into biological processes and vice versa. So we can talk of bio-socio-material entanglements.

Ideas like Keller's are often discussed under the heading of the Extended Evolutionary Synthesis.[21] These new approaches constitute a strong corrective reaction to the neo-Darwinian, gene-centered synthesis. Biology is integrated into complex webs of social, material, and practical entanglements.

The result is that accumulated traditions can have biological components. For example, the Neolithic East Mound at Çatalhöyük in Turkey was inhabited by up to eight thousand people for over one thousand years.[22] That very stable community passed down a great deal of accumulated information: how to build houses, how to plaster them, how to manage waste, how to bury the dead. This cultural information was embedded in material processes as people worked with clay to prevent house walls collapsing and found technological solutions to provide sufficient food for a population that over time increased in size and density. The layouts of houses and management of resources were governed by strong social and cultural rules, but these codes also incorporated practical solutions to the disease risks encountered in a densely packed settlement. Bright Zhou describes the many layers of marl plaster on the house walls, the careful replastering of floors with marl clay after human burial beneath them, the assiduous "symbolic" separation of clean and dirty floors in houses, and the sweeping of refuse into open areas or middens where there was much human, sheep, and dog fecal material but that were frequently burned or covered over with clay.[23] He notes that the marl clays used at Çatalhöyük were high in calcium carbonate, which would have created an alkaline environment that inhibited fungal, mold, and bacterial growth. We know little about the emergence of gluten tolerance among early farmers,[24] but it is likely that epigenetic and genetic changes were taking place as humans invested more in gluten-rich diets. For some people, impaired gluten absorption would have contributed to iron and nutritional deficiencies and to the disease and material interactions shown in figure 3.1. Thus, humans, material things, bacteria, and disease were caught up in a heterogeneous mix that developed over time. The accumulation of components had biological, social, and material aspects.

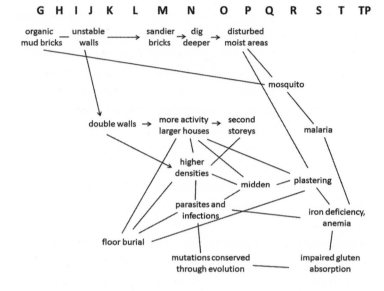

Figure 3.1. The entanglements of biological, social, and material things through the seventh millennium BCE at Çatalhöyük in Turkey. The earliest occupation level, G, is to the left, and the latest occupation, TP, is to the right. Source: Author.

And what was there constrained and directed what could be done next. Plastering and midden practices mitigated the spread of disease, but they involved considerable labor (digging marl clays, mixing organic materials into foundation layers, spreading it on walls, burnishing). Humans got caught up in the continual attention to walls, floors, burials, and middens. We presume that the inhabitants of Çatalhöyük did not have a theory of disease similar to our own, but they may have noticed that deviance from the ritually sanctioned separation between clean and dirty areas of the house led to more illness. Symbolic codes and material practices may have become caught up in each other. At a larger scale, although humans at Çatalhöyük had worked out ways to mitigate the disease risks of living in large, stable settlements, the long-term result was an increased prevalence of disease in urban settings.[25] Humans were drawn into more intensive agricultural systems and larger urban settlements, but they also found themselves entrapped in a wide range of health risks.

The Extended Evolutionary Synthesis is important because it explains how biology, materials, and society get caught up in each other and bring about the emergence of accumulated traditions. Rather than organisms responding to environments and niches, there is a cocreation in which humans, things, plants, animals, and social relations—all these intra-actants in Karen Barad's term—coproduce change.[26] The process of change is undoubtedly cumulative as solutions to problems are based on existing practices. But there seems to be no reason to argue that the juggernaut of interactions should move in any particular direction. Evolution is cumulative, and it is constrained, but that does not mean we can predict where it will go.

Unless, that is, we argue that something in this mix is pulling the whole juggernaut along. Through most of earth history, that does not seem to have been the case, and simpler organisms at times have been selected for in specific environments. But since the emergence of humans and their heavy dependence on material things, there has been a process pulling the juggernaut toward more stuff.

A juggernaut that includes humans works differently from one that does not. This is less a qualitative difference than a difference in degree. And it may seem blindingly obvious—of course, humans make a difference! But in fact much recent theory tries to explain human social and cultural development in biological terms, or using analogies to biological evolution. I argue that the human dependence on things makes human evolutionary change different from biological evolution—and that this difference explains the directionality of human evolution. I want to start with the example of dams: those produced by beavers and those produced by humans.

Beaver Dams and Human Dams

Beavers build dams, and so do humans, and this fact is often used to argue that humans are not the only animals that engage in ecosystem engineering. Beaver ecosystem engineering is remarkably complex and goes well beyond building dams. Beaver dams, made from wood, mud, stone, and earth, create the ponds in which the beavers live. They also build lodges, normally in the center of the pond, and they inhabit bank burrows around its periphery. The raised water level extends the pond into surrounding woodland and vegetation, allowing the beavers to create water channels in which they swim to collect wood that can be floated back along the channels to the lodge and dam, all the time protected from predators. Some exceptionally large North American beaver dams have been recorded. One in Montana was seven hundred meters long, and one in Wyoming exceeded a height of five meters. Beavers can also create systems of dams—in one case including twenty-four dams along a thirteen-hundred-meter section of stream. The dams can last for decades, creating a series of "monuments" in the landscape.[27]

The beaver dams cause a buildup of sediment, slowing stream velocities, and they change the landscape from a fluvial to a wetlands environment. Depending on local conditions, they may change hydrology, temperature, and water and soil chemistry.[28] Overall, beavers tend to increase habitat and species diversity at the landscape scale. "As the habitat is altered, interactions amongst co-habiting species may change. For example, brown trout or brook trout (charr) may become dominant over Atlantic salmon."[29] In addition, both inhabited and abandoned or collapsed dams have a major impact on local vegetation and the succession of plants as well as on a wide range of animal, reptile, and bird species.

The dams require "continuous supervision and perpetual labor," wrote Lewis Henry Morgan.[30] The effects of water percolation, seepage, settling, and decay demand careful vigilance and repair. In fall, the beavers add a new supply of materials to the dams to compensate for decay, using wood cuttings from the previous autumn that have been stripped of bark for food and purposely laid aside for use in the repair. Dams may fail for a number of reasons, including floods, animals burrowing through them, and the collapse of dams upstream. Many humans look on beavers as pests and try to destroy their dams. But even when a dam is dynamited, the beaver family works hard and fast to repair it. Sometimes, with a fairly new dam, the beaver colony will move but only to the next strategic location up or down the river or ditch. Whether beavers repair a broken dam depends on the season and on the size of the break.[31]

A more effective way to eradicate beavers is to kill or destroy young willow or cottonwood trees near their dams.[32] These trees provide the beavers' basic food and construction material. They can repair a dam but they cannot replant trees. This might seem an obvious point, but it shows that there is a limit to how far beavers will go down the chains of consequences and put problems right. Humans follow further along the chains of consequences.

China's Three Gorges Dam on the Yangtze River is the largest dam made by humans. It is nearly one and a half miles across, produces 18,200 megawatts of hydropower, cost between $25 billion and $40 billion, and displaced 1.6 million people. It provides considerable benefits in electricity generation, irrigation, and flood control, as well as to the pocketbooks of politicians, bureaucrats, and construction companies. But the chains of consequences are often very negative. For example, the Three Gorges reservoir was in 2008 described as a cesspool because of the hundreds of factories, mines, and waste dumps that were inundated by its creation.[33] The rising reservoir set off landslides, forcing communities to relocate. Even at the mouth of the Yangtze, hundreds of miles downstream, saltwater moved up the river as the flow of freshwater declined, and as a result tidal wetlands near the mouth deteriorated.

The impacts have also been cultural. Many archaeological and cultural sites were threatened by the Three Gorges inundation, but archaeologists were able to investigate some 1,087 sites throughout the valley, gaining information about people who lived there as much as 2 million years ago, as well as for nearly every era of habitation since. But untold information was lost to the rising water levels.[34] Saving the more important sites involved costly interventions. The Baiheliang Stone, the best preserved ancient hydrologic station in the world, was saved by constructing an underwater museum around it. Shibaozhai, an ancient Buddhist temple built by the Ming emperor Wan Li, is now an island in the center of the new lake, surrounded by a concrete dike. Some structures were moved altogether, such as the Zheng Fei Temple, which was taken down one brick at a time and reassembled at a higher elevation.

The damming of the Nile brought similar consequences. From the time of the earliest agriculture in the Nile valley, the annual floods have replenished fields, and ditches and dikes were built to control the spread of water over the floodplain. The Middle Kingdom Twelfth Dynasty (1991–1786 BCE) introduced large irrigation schemes that successive governments maintained.[35] Recent attempts at dam building began in 1889 at Aswan. The numerous consequences of the late-nineteenth- and early-twentieth-century dams included the spread of malaria in the 1940s, shortages of chemical fertilizer, the growth of pondweed, silt accumulation, and evaporation.[36] The Portland cement used for the dam led to leakage and erosion.

Egypt's president Gamal Abdel Nasser envisioned a still larger dam, and the Aswan High Dam was constructed between 1960 and 1971 at a cost of US$1 billion. According to Fekri Hassan, the benefits were providing perennial irrigation, an increase in the area of arable land, increased access to drinking water all over Egypt, improvement in river navigation, and flood control.[37] The High Dam Lake became a major source of fish, and the electric power it provided was essential for the country's industrial development. But such a large area was flooded that one hundred thousand people had to be relocated. There has been coastline erosion and increased soil salinity caused by the year-round increase in the water table, requiring a subsurface drainage system over an enormous acreage. The reservoir has gradually silted up, reducing its water storage capacity. This is being dealt with by new projects to build dams higher up the Nile. Finally, the dam construction caused aquatic weeds to grow faster in the canals, and this had to be dealt with using a variety of biological and mechanical methods.

Thus we see humans interfering in river flow, leading to multiple environmental consequences that then have to be addressed. But remarkably, humans are also concerned about the monuments and sites of past societies. They could simply let these remains be flooded, yet for various reasons, they feel the sites have to be saved from destruction. Although all human societies have a concern with the past, since the eighteenth-century CE, humans in some countries have grown increasingly sensitive toward heritage and have come together around saving it. There is no need for them to have become entangled in the past in this way, but a series of factors have led to a rise in awareness of past monuments, sites, and cultural artifacts.[38] Egypt and, of course, the Nile have long been the focus of occupation and monument building, making it inevitable that the Aswan High Dam would have a major impact on archaeology. The Nubian Monuments Campaign, the first high-profile collaborative international rescue effort, was mounted with the aim of salvaging the sites. One of the main outcomes of this campaign was the valorization of what came to be known as "world heritage" in the World Heritage Convention of 1972 and the establishment of a UNESCO World Heritage Center in 1992 entrusted with the mission of safeguarding the cultural heritage of humankind. One aspect of this campaign was that Egypt suffered a major loss of cultural heritage as some of the finds and temples were taken to major museums in Europe and the United States. It was decided to lift some of the monuments to higher ground—the most famous of these being the temple of Abu Simbel, which was cut up into large blocks, relocated, and reassembled. A positive consequence has been increases in tourism to the monuments of the Nile, including Abu Simbel. A more negative consequence has been that hydrographic changes such as the height of the water table have affected the preservation of monuments all along the Nile.[39]

The emergence of UNESCO and the World Heritage Convention can be seen as responses to the consequences of human intervention in nature and large-scale construction. The management of these consequences has led to massive bureaucracies that regulate and oversee heritage management at international, national, regional, and local levels. This is a peculiarly human concern, specific to particular historical periods and regions. As we have seen, beavers produce monuments that have long duration that then collapse and decay. But beavers, perhaps sensibly, do not worry about protecting their ancient dams and lodges.

Beavers construct things that act as a selective niche for themselves and other organisms, and humans do the same thing. But there is a major difference. We have seen that beavers deal with the consequences of their interventions in nature. I gave examples above of how they are caught in an endless cycle of repair to dams. Lewis Henry Morgan gives examples of where one dam creates a river flow problem that beavers address by building secondary dams.[40] We saw this solution also used for the Aswan High Dam. But there is a limit to how far down the chain of consequences the beaver will go. For example, beaver dams may impede the movement and spawning of fish. Human dams do the same thing, but humans often find ways to deal with the problem by constructing lifts, "ladders," passes, or steps so that migrating fish can get past dams. Humans get drawn into further investments in order to deal with the consequences of their actions.

HUMANS AND THEIR CONSEQUENCES

One of my reasons for using the example of the beaver is that an early source of research on the beaver is *The American Beaver and His Works* by Lewis Henry Morgan, an American anthropologist and social theorist whose ideas about social organization and the role of material culture and technology in social change influenced Karl Marx and Friedrich Engels. Morgan drew parallels between the engineering knowledge of people and of beavers, and in a book dedicated to the role of things in human evolution, it seems appropriate to ask what Morgan's careful study of the behavior of beavers taught him about the differences between humans and animals.

After observing beavers' intelligent responses to problems such as the decay and destruction of dams, unpredictable water flows, and ecological change, Morgan argued that beavers do not work simply by "instinct" but that they have a "thinking self-conscious principle, the same in kind that man possesses, but feebler in degree." For example, describing the construction of canals from the pond to the higher ground on which hardwoods are found, he shows that this effort involves forethought and planning, and awareness of consequences of taking one action over another. His book focuses on "the works" of the beaver, and he makes clear that he sees the beaver as a "lower animal," a "mute," but one capable of remarkably complex engineering. But the beaver's skills, learned and passed down through generations, are limited by its intelligence. It can deal with only a limited range of consequences.[41]

Humans are quintessentially tool users and makers (*Homo faber*). But they also have a large and complex brain and nervous system (*Homo sapiens*). As a result they see consequences everywhere, and they try to invent fixes for the problems they see emerging. In the case of the Aswan High Dam, they dealt with the problem of increased soil salinity by introducing a subsurface drainage system. The problem of aquatic weeds has been addressed with biological and mechanical interventions. Mechanisms have been found so that fish can migrate upstream past dams. The destruction of archaeological sites has been dealt with by lifting sites to higher ground, conducting surveys prior to flooding, and establishing international heritage agencies dedicated to the sites' protection. Humans get drawn into managing the consequences of their actions, and this entanglement brings a tendency for general directional change.

We have seen that there seems to be no agreed-upon account of a general directionality in biological evolution. While there is good evidence for specific directionalities as organisms are selected for that are better adapted to a particular environment or niche, there is no generally accepted theory that would explain a general directionality, even if ways could be found to define that directionality (in terms of complexity, for example). There are also problems with the idea of biological adaptation to environments and niches; a more valid approach describes adaptation as involving multiple agents, ranging from the biological to the social, material, chemical, and physical. The overall sets of bio-socio-material processes are cumulative in that they build on what is already there, and in genetic terms, genotypes are legacies of evolutionary paths taken long ago. But this cumulative character of evolutionary processes does not produce an overall direction.

While it seems, then, that biology does not provide an answer to the question of why human evolution is directional, in this chapter we have been able to isolate some aspects of a difference between biological evolution generally and human evolution specifically. As the comparison of beaver and human dam building makes clear, human engagement with and dependence on things and their greater attention to problem solving distinguishes them from other species. Humans get drawn into the consequences of their actions in ways that other species do not. They follow further down the chains of interactions and try to solve emerging problems. We shall see that this characteristic is what leads humans into a directionality perhaps not seen elsewhere in biological evolution.

Humans and Things

T HE NOTION that things and environments made by and engineered by organisms play a role in evolution has been argued very effectively by niche construction theory. This theory describes how humans construct niches (of cultural and ecological information as well as physical habitats) that affect the selection pressures on themselves and other organisms.[1] It explores complex and dynamic relations between organisms and environments and allows study of human-made transformations that have created novel ecosystems. Niche construction is largely interpreted as modification of the environment—ecosystem engineering. In such terms it has a meaning close to the term "artifact," albeit largely with regard to modified landscapes. However, rather than deriving a theoretical focus from debates about artifacts, things, and the dependencies within intersecting operational chains, a theoretical framework is used that largely derives from evolutionary theory. The latter is relevant for human evolution, but as we saw in the example of dam building, human relationships with things differ from those of other organisms in important ways. In any case, evolutionary selective pressures act far too slowly to explain the vast majority of human relationships with things. So in order to understand how humans and things have changed together through time, we need to turn to theories that derive from sources other than biological evolution.

Let's start with a very basic question: What is a thing? The answer is rather more complex than we might hope, but the complexity is what makes things interesting and key to the theory of entanglement. Take the example of a wheel. I said in chapter 1 that it is very difficult to know where and when the wheel originated. Part of the problem is definitional—what exactly is a wheel? The wheel is often taken as the classic example of an invention that has changed the world. Much has been written about the origins of the wheel, but when you think hard enough about it, you realize that different definitions lead to different origin narratives. We might define it very narrowly as a thin circular object with a hole in the middle. Even that raises difficulties: Some wheels are thick and not exactly circular, and some do not have holes in the middle. But a wheel thus defined cannot function without an axle, and an axle needs a frame or vehicle to hold it in place; so where is the boundary between the wheel and the wagon? What about the person or cow or horse or internal combustion engine that drives the wagon? They are necessary for the wheel to function. A full understanding needs to explore all of these unbounded entanglements. Talking about wheels is very difficult without also talking about roads, and therefore the construction of roads and their maintenance by the state or other bodies. In modern times, distant events can have quantum effects. For example, events in Saudi Arabia and the Persian Gulf affect the price of oil and thus the numbers of wheeled vehicles on the roads in California. "Wheelness" is distributed, displaced, deferred, and dispersed. Thingness is a dispersal and a making of connections.

The wagon wheel brought together and was made possible by

many other things and had innumerable consequences. As shown in figure 1.5, moving heavy objects on wheels may have derived from the use of rollers, and the idea of the wheel was already present in the earlier use of the bow drill, the spindle whorl, and perhaps also the potter's wheel. Because use of a wheel for transport depends on the existence of an axle and a frame for the axle, such as a wagon, the wheel depends on fairly complex carpentry and carpentry tools: Making a large, flat, circular wheel is a difficult and skilled process. Whether it is the axle and wheel or just the wheel that rotates, some lubricant is needed, initially animal fat. The animal that is to pull the wagon, such as an ox or horse, needs to be attached to it by some system such as a yoke. The training and yoking of cattle in the European Neolithic was also linked to their use to draw the plow. So the wheel brings many things together; the conditions of possibility for the wheel are dispersed, extensive, and seamless.

No thing is a thing unto itself. The etymology of the Anglo-Saxon word "thing" refers to an assembly or bringing together in a "ting." As Martin Heidegger noted, all things draw together.[2] He talked in terms of a jug, but we can see the same process with the pots made in the Neolithic of the Middle East. For example, the earliest cooking pots at Çatalhöyük brought together existing ideas and technologies such as fired clay (earlier used for figurines), containers (baskets and wooden bowls), water, fire, hearths, and the cooking of food (figure 4.1). The making of pots also had consequences, such as more efficient cooking, that allowed more activities to take place in the house.

Many of the connections between things are operational chains: They involve sequences of actions that lead to some end result. Archaeologists are used to studying these upstream-downstream processes, from procurement to manufacture to use to maintenance, repair, and discarding (figure 4.2). These sequences are then cross-cut by the sequences necessary to produce the tools to make things in the original sequence. The operational sequence for the production of pottery is thus cross-cut by the operational sequence for the wheels on which the pots are formed, the kilns to fire the pottery, and so on, in very complex webs of connections. As another example, the sickle was involved in complex operational chains associated with the domestication of plants (figure 4.3).

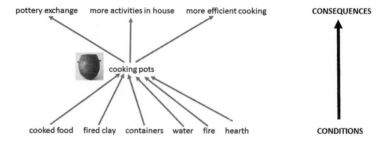

Figure 4.1. Some of the conditions and consequences of the introduction of cooking pottery during the seventh millennium BCE at Çatalhöyük in Turkey. Source: Author.

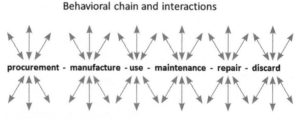

Figure 4.2. The behavioral or operational chain from the procurement of a raw material to the production and use of a thing to its discarding, and the cross-cutting interactions with all the tools necessary at each stage. Source: Author.

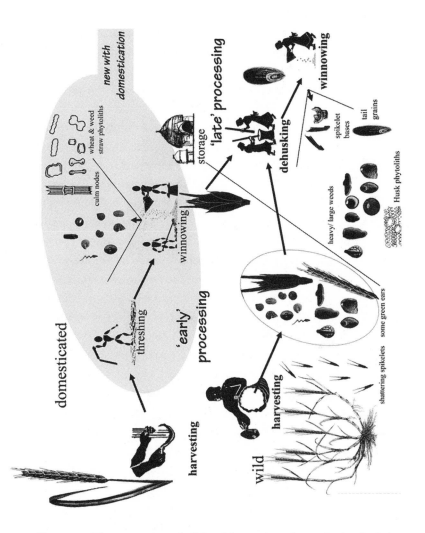

Figure 4.3. The processing of wild and domesticated cereals; the shaded area indicates the extra labor needed for domesticated plants.
Source: D. Fuller, "Contrasting Patterns in Crop Domestication and Domestication Rates." Reprinted with permission.

There are, of course, many forms of connection between things, including the biological, social, and ideational. I have argued elsewhere that varied types of things, including concepts, institutions, and even humans, need to be included in operational sequences.[3] Institutions and notions of class or political inequality affect how operational sequences play out. We saw in chapter 2 that the idea of progress was a component in the dispersed practices of capitalism, colonialism, and empire that dominated much of the world in the nineteenth century CE. Ideas can certainly become things that have complex interactions with man-made material things. Humans build castles in the air and on the ground, and both create entanglements. Henri Bergson wrote, "We shall see that the human intellect feels at home among inanimate objects, more especially among solids, where our action finds its fulcrum and our industry its tools; that our concepts have been formed in the model of solids; that our logic is, pre-eminently, the logic of solids."[4] In this general sense, concepts and things have an apparent solidity and fixity. But, again, the thingness of things, including ideas and institutions, consists in the interconnections, the lack of independence, the relationality of entities. Things cannot be separated from the *idea* of things and vice versa: As we saw with the wheel, when different definitions or meanings are brought into play, the object of study starts to shift. The wheel-thing depends for its existence on the "idea of a wheel"–thing, as much as the "idea of a wheel"–thing depends on material things and their interactions.

While in this book I am mainly talking about human-made material things in their relationships with each other and with humans, it is important to recognize that ideas, institutions and even humans are themselves things that participate in complex interconnected webs. I focus on human-made things—artifacts—

because they draw humans in a general evolutionary direction. While ideas and goals play an important part in orienting human agency in specific directions, the separate and independent agency of things produces the overall directionality. Made things are both part of humans and separate from them, and this duality is what creates directional change. Ideas, institutions, and other humans also become "objectified" and "reified," separated from humans, and thus have quasi-independent agency. But we will see in chapters 6 and 7 that general directionality is primarily the product of human interactions with material things they have made.

Spinning Cotton

For now I simply want to make the point that the thingness of things involves the drawing together of many different strands (including ideas and institutions). Let us return to the example of cotton, which we looked at in chapter 1.[5] The expansion of spinning technology from simple spindle whorls to massive modern spinning machines is an example of the increased human investment in material things. But it was overly simplistic to illustrate in figure 1.4 the linear sequence of the increasing complexity of cotton spinning machines (or indeed of wheels or harvesting machines in figures 1.3 and 1.5). Those diagrams "objectified" the objects, separating them from the conditions and consequences of their existence. If we see them as "things," on the other hand, we can explore their interconnections and dependencies. In figure 4.4 I have tried, for the case of cotton, to point toward this wider cone of linkages.

The story of cotton, from a long, slow start to a massive contemporary global industry, is very much tied to the nature of the crop itself. Cotton involves several stages of labor-intensive work:

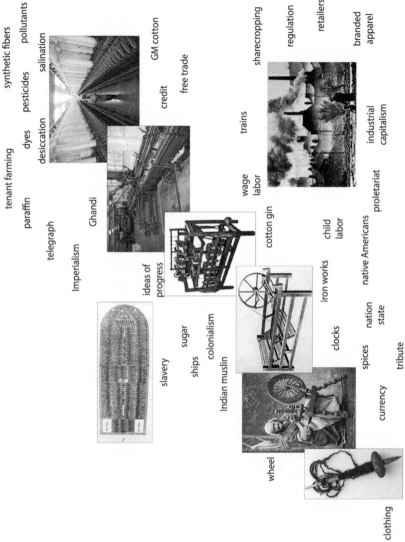

Figure 4.4. The development through time of spinning technologies as in Figure 1.4, with some illustration of the expanding cone of entanglements. Source: Author.

picking, separating the fibers from the seed, spinning, and weaving. It took a very long time, and very particular circumstances, for the conditions to be met for cotton's full affordances to be realized. In about 3000 BCE, the first cotton threads started to be used in India and Peru. There was then a long dormant period when cotton played a small role as clothing, tribute, and currency. Partly because of the difficulty processing cotton and manufacturing cotton cloth, Europeans depended on wool and flax into the late medieval period. But then production in Italy using cotton imported from the Levant led to the start of a cotton industry that later spread to Germany. Production was still based largely on rural labor. It was Europeans' willingness to inject capital and force into global cotton networks through armed trade that allowed the various stages in cotton manufacturing to be distributed around the globe. In the sixteenth and seventeenth centuries, the Portuguese, French, Dutch, and especially the British through the British East India Company began to set up maritime trade networks in which cotton textiles were purchased in India, traded for spices in Southeast Asia, and also brought to Europe, where they were consumed or sent to Africa to pay for slaves to work the cotton plantations then starting in the Americas (building on the model of sugar plantations). This was a true entanglement of a multicontinent system.

Asia, the Americas, Africa, and Europe became tied together in a complex commercial web. Through the seventeenth and eighteenth centuries, the cotton industry remained on a fairly small scale in England, but increased demand and the increased volume of the slave trade allowed the gradual growth of factories in England. Then, in Manchester in 1784, something new started. A new thing, the water frame, used waterwheels to spin yarn from Caribbean cotton, exploiting orphaned children and local workers. There were, of course, many causes of the Industrial Revolution and the expansion of cotton production in England in the 1780s, but certainly one crucial factor was the invention of the water frame.

Spinning had been done using spinning jennies in the home. The water frame allowed larger-scale production, as did the ensuing mules (so called because, like actual mules, they were hybrids of the water frame and the spinning jenny) that were the basis for the contemporary gigantic machines. The effects of these early machines were considerable. Before 1780 the relatively high wages paid to workers in England meant that manufacturers had difficulty competing with Indian output. The new technologies changed the equation by mechanizing and speeding up production. Cotton could now be produced cheaply and on an increasingly massive scale. The pace of labor changed, wage labor appeared, and other industries such as the railroad network and iron production were also affected. Clearly it is too much to say that the new spinning machines *caused* the Industrial Revolution, but they played a key part in the emergence of an industrial capitalism involving new economic, social, and political institutions.

The machines pulled vast numbers of humans into the cities, and into terrible conditions. Why did people move into towns and take on dreary, tedious work over which they had no control? Partly it was the result of a long process by which land had been enclosed and taken over by wealthy landlords, but partly it was because the factories and mechanized cotton production had made cotton production much cheaper, so that people working at home could no longer compete. Machines forced people into factories and wage labor. The machines also created great wealth—and the wealthy mill owners gained political power, allowing them to exploit workers further, for example, by passing laws allowing the employment of children. But the workers also gained power, because someone has to operate the machines. So emerged the conjunction of power and class in modern democracies. The creation of the modern world was in part pulled along by the machine.

Working in factories was bad for workers' health, and in 1753 there were attacks on cotton machines. In 1811 and 1812, steam looms were attacked in the English town of Stockport, outside Manchester. Machine breaking was declared a capital crime (the state intervened on behalf of mill owners), and people were hanged. Mill owners came to depend on the state to suppress upheaval. Workers nevertheless collectivized and fought for higher wages, shorter working hours, and better working conditions. Spinners formed unions in the late eighteenth century. Much of the British working class gained the vote in 1867.

In the United States, the new class of factory owners came into conflict with the slave-owning farmers in the South. Many cotton industrialists in Britain were concerned that they had become too dependent on the American South for raw cotton, a crop that had become very productive for the United States. Several factors accounted for the South's dominance of the cotton trade: its use of slave labor, the vast tracts of land it had acquired through the killing or forced emigration of Native Americans, the invention of the cotton gin and its spread throughout the region, and the development of a financial system that made credit readily available. The Civil War resulted in part from the conflict between the Southern need for slavery and the sensibilities of Northern industrialists.

The Civil War halted the flow of cheap raw cotton into British factories, and production failed to fully recover after slavery was made illegal in the United States. British manufacturers turned to India, and the production of cotton shifted to the Global South. In the latter half of the nineteenth century, US production decreased dramatically and became heavily subsidized by federal agencies. Today, throughout India, China, West Africa, and other parts of the Global South, about 350 million people are involved in global production—growing, transporting, ginning, warehousing, spinning, weaving, and stitching. Some nations force farmers to produce cotton despite the devastating environmental and financial consequences. For example, in Uzbekistan, cotton farming has caused desiccation and salination of the landscape. Many farmers are forced to sow genetically modified cotton, which is more expensive to buy and maintain but also more productive, pushing cotton prices down. Farmers can thus get locked into a cycle of debt and forced production. In India in 2005, after a poor growing season, hundreds of heavily indebted farmers of genetically modified cotton committed suicide by drinking their own pesticides. Each year, 2 billion cotton T-shirts are sold worldwide; their manufacture not only uses huge amounts of water and energy but also releases by-products of starch, paraffin, dyes, pesticides, and other pollutants into the air and soil. The social and environmental costs of spinning cotton have not disappeared: They have been transformed from the horrors of slavery and child labor to chemical pollution and financial indenture.

The global cotton industry is now driven not by manufacturers and producers but by huge retailers and branded apparel sellers like Walmart and Carrefour. These immense corporations, which can easily switch from one source to another, determine production and distribution in their own interests. The result is enormous global inequality. Between 1995 and 2010, the owners of cotton-growing businesses in the United States collectively received government subsidies of more than $35 billion. Meanwhile, a cotton grower in Benin in 2014 made a dollar a day or less.

We see then that the evolution of machines from spinning jennies to water frames and mules to giant factories created an expanding outer cone of conditions and consequences (see figure 4.4). In the case of cotton spinning, this outer cone involved people, material things, institutions, and ideas, including sugar, tobacco, trains, clocks, the telegraph, wool, flax, guns, spices, iron, pollutants, ships, tribute, clothing, unions, slaves, children, Native Americans, trading companies, entrepreneurs, retailers, debt, machines, wage labor, the movement into towns and the emergence of an industrial proletariat, industrial capitalism, enclosure, the nation-state, governments, colonialism, and much more.

We cannot assume that the results of this growth are coherent or in tune with each other. In fact, things can evolve in contradiction or opposition. Sven Beckert describes how cotton "brought seeming opposites together": slavery and free labor, states and markets, colonialism and free trade, industrialization and deindustrialization, plantation and factory, colonizers and colonized.[6] In the seventeenth and eighteenth centuries there were contradictions between, on the one hand, the rule of law and individual rights within European nation-states, and on the other hand, the use of force by private companies to impose suffering and subjugation in other parts of the world. A more basic contradiction was that nation states did not have the bureaucratic, legislative, or policing resources to manage the global trade, so they left it to private companies and armies. In the mid-nineteenth century, slavery in the cotton-growing US South came into conflict with a rising class of industrial owners whose concerns for the human condition led to a disavowal of slavery. In more recent times, many of us profess concern about environmental and social issues and yet wear T-shirts that sustain a global trade involving massive inequalities, hardship, and environmental pollution.

We can already glimpse a point that I discuss more fully later in this book: that the evolution of things like cotton-spinning machines is pulled along by these outer entanglements and their contradictions. Most clearly, for many millennia cotton production remained at a low technological level. This was partly because compared to wool or linen, cotton was relatively hard to process. It was the particular conjunction of slavery, machines, international trade, and colonialism in the late eighteenth century that allowed humans to realize cotton's potential. As an example of how contradictions in outer entanglements can produce evolutionary change in things, in the late eighteenth century the relatively high wages paid to British workers came into contradiction with the low wages and low-cost cotton production in India, leading to the invention of the water frame. In industrial Britain, people living in rural areas produced cotton that could not compete with that produced by machines in towns; this conflict led to migration into towns and the formation of an urban working class that needed work, stimulating the expansion of spinning production and the invention of still larger and more efficient machines.

The Example of Fire

The "thingness" of cotton and cotton-spinning technologies is enormously dispersed and complicated. The process of spinning cotton starts with a tangled mass of fibers of different length. There is then first the drawing out or drafting and laying side by side of the desired number of fibers, then the twisting of these fibers so that they become bound together as yarn. But the fibers extend out and draw in numerous other things in the outer cones of entanglement. The whole entanglement is a tangled mass that is very difficult to sort out. Other examples of equally complex global entanglements are Anna Tsing's discussion of the matsutake mushroom and Timothy Mitchell's account of the links between democracy and fossil fuels.[7]

At least initially, the making of fire looks much simpler. A diagram of the things brought together by the making of fire in early prehistory would be relatively straightforward. For a fire made by rubbing two sticks together, we might start with a piece of hard wood rotated by human hands in a piece of soft wood, some dry grass, and human breath gently blowing on the spark as it lights the grass.

Over the long term, we can track the consequences of the human production and control of fire, from the early use of fire to cook food; to its use to make pottery, then metals, glass, lime and concrete, steam trains, internal combustion engines, gas lighting, and oil-fired central heating; as well as to the use and demand for fuels and sources of heat from wood and dung to coal, oil, electricity, and nuclear energy. Clearly this is another example of a long-term trend toward greater human engagement with things and the consequences of things.

But in its early use, fire gives us a relatively simple context in which to explore the thingness of things. It is difficult for archaeologists to distinguish accidental from intentional fire, and many of the claims for early use of fire by humans are contested. Frances Burton suggests that the use of fire started around 6 million years ago at about the time of the divergence between chimps and humans, and that there was gradually closer association with fire until it was manufactured by modern humans. Richard

Wrangham, on the other hand, suggests that the earliest users of fire were probably *Homo erectus* about 2 million years ago, but he notes that the first human use of fire of which we are reasonably certain is at 790,000 BP at Gesher Benot Ya'agov in Israel. Later examples occur at Beeches Pit in England at 400,000 BP, and at the same time in Germany at Schöningen. From that point the unambiguous cases of human use of fire increase substantively in association with Neanderthals and *Homo sapiens*.[8]

A number of researchers have argued that fire had numerous implications for the early evolution of hominins. For example, Burton, who accepts evidence for very early use of fire, suggests that "proximity to fire stretched out the period of light and irreparably altered hormonal cycles that are dependent on light and darkness." The physiology of reproduction changed. Also, "firelight, by inhibiting melatonin, enhanced memory formation."[9] But it is Wrangham who has offered the fullest argument for the importance of fire for human evolution, with implications that are as much biological as social and economic.

Wrangham notes that cooking increases the amount of energy our bodies can obtain from food. Humans have relatively small digestive systems compared to our cousin apes, but in these smaller guts they can process cooked food efficiently. Compared to chimpanzees and other apes, humans have small mouths, weak jaws, small teeth, all attributed to cooking. Cooking creates softer food, higher caloric density, low fiber content, and high digestibility.

Wrangham cites the work of Robin Dunbar showing that primates with bigger brains or more of a neocortex live in larger groups, have a greater number of close social relationships, and use coalitions more effectively than those with smaller brains.[10] Diet is a key driver of human brain growth, which takes a lot of energy. Primates with smaller guts tend to have larger brains, thus "primates that spend less energy fueling their intestines can afford to power more brain tissue."[11] Leslie Aiello and Peter Wheeler suggest that increases in human brain size would be associated with increases in diet quality, and Wrangham argues that the improvement in diet quality came from cooking. Particularly in the Upper Paleolithic, technological advances such as better hearths, ovens, and containers would have increased the efficiency of cooking, enhanced diets, and supported larger brains and larger coalitions.

Gathering around a campfire would have selected for humans with greater tolerance and desire for each other's company. Sharing the source of fire would also have contributed to sociable behavior in environments where fire-starting was not so easy. In many societies, the hearth and fire are the center of social and family life and have symbolic and spiritual importance. Wrangham lists a number of other implications of the use of fire, including deterring predators. But his main focus is on cooking. For example, foods soften when cooked, which meant that humans did not have to spend so much time chewing and were freed up for other tasks that led to the sexual division of labor. The men could spend more time hunting, and the women needed the men to protect their food. Also, cooked foods, being soft, would have allowed early weaning, and so larger families.

Wrangham concludes that "cooking increased the value of our food. It changed our bodies, our brains, our use of time, and our social lives" but also made us "dependent on fuel."[12] Thus the small number of immediate entanglements involved in lighting a fire very quickly expanded out into a wider sphere of entanglements. These again ranged from the cultural and social to the economic, ideological, and biological. And they very much pulled the use of fire along. Once they had evolved smaller guts and larger brains, humans became very dependent on cooking. Once they migrated into areas of extreme cold and ice, their existence depended on fire. One of the most fascinating items of equipment found with Ötzi the Iceman who died in the Alps about five thousand years ago, was a pouch containing a kit that included a tree fungus used for tinder, and iron pyrites and flints for making sparks.[13] He also carried a birch bark container holding charcoal and maple leaves, which may have been for transporting embers wrapped in leaves. The care of and maintenance of fire, and in particular the pursuit of fuel, were important components of prehistoric humans' lives. Fire kept people warm, cooked food, cleared landscapes, and later transformed clay and metals and provided transport and numerous forms of energy. Humans became very dependent on it, but they also had to deal with its negative effects. The Anthropocene, a term that has come to denote the period in which carbon emissions from industrial activity have had a severe impact on global environments, is often thought to have started in the nineteenth and twentieth centuries with the rise of large-scale industries. But one might also argue that the Anthropocene began with the earliest human use of fire.[14] Over time, the fueling of fire led to massive extraction of wood forests, peat, coal, and oil, and to major environmental change. We cannot do without fire, but our need draws us in to work and care.

Why We See Objects and Not Things

If I have belabored the point about the interconnectedness of things, it is because we often do not see these connections. When we take the strings of lights off the tree after Christmas, they tend to get tangled. If our string is very old, sometimes one of the bulbs fails, so the whole string will not work. For various reasons, people in the United States throw a lot of them away every year. Where do these discarded lights go? Many go to the southern Chinese town of Shijiao, whose factories import and process 2.2 million pounds of discarded Christmas tree lights every year. Cheap labor and low environmental standards have made the town an important center for the recycling of these lights. Until recently, many factories in Shijiao burned the lights to melt the plastic and recycle the copper wire, releasing toxic fumes. Today, as Adam Minter tells us, they use a cleaner method.[15] When the Chinese started to buy cars in large numbers, the price of oil went up, as did plastics made from oil. Recovered plastic thus became valuable as an alternative. Instead of burning the plastic off the lights' copper wires, people figured out a way to strip it off and reuse it. The lights are tossed into shredders, and workers then separate the resulting material on vibrating tables spread with water. The plastic they recover is of a good enough grade to be made into slipper soles, and the copper is used in plumbing, power cords, and smartphones.

The making and recycling of Christmas lights provides employment for people worldwide. They are part of a heterogeneous network of religion, commerce, trade, and production (as well as slippers and plumbing) with global reach. We could do without them, they use a lot of resources, and their discard can cause pollution. Yet it is in everyone's interests to keep using them. Christmas tree lights are one way that economically developed countries export their junk, and the hard and dirty labor associated with them, to other countries. The people who have come to depend on Christmas tree lights in various ways do not want to cease their production, use, and recycling despite the pollution, energy "waste," and perpetuation of global inequalities.

One reason we do not "see" all the pollution, low-paid labor, and appalling work conditions is that many of us remain distant from it. As we innocently stand on our stepstools putting the lights on the tree, we are oblivious to the planetary entanglements and entrapments we create. China and other rising nations make massive profits from our recyclables—not just Christmas tree lights, but televisions, cars, mobile phones, paper, cardboard, and much else. We recycle, but it is almost as if we want to convince ourselves that our headlong rush to stuff has no implications for our entanglement with the planet. For example, we give our new digital technologies insubstantial names like "air," "cloud," and "web" even though they are based on buildings full of wires, enormous energy consumption, cheap labor, and toxic production and recycling processes. When we count the wireless connections, data usage, and battery charging, the average iPhone uses about 361 kilowatt-hours (KWh) each year.[16] A medium-sized refrigerator with an Energy Star rating uses only about 322 KWh a year. The problem is not the phone itself but all the systems that run continuously to support it: the computers and servers running twenty-four hours a day, seven days a week; the air-conditioning systems needed to keep the servers cool; the manufacturing centers to build the devices; and the nonstop electricity needed to power the broadband networks. Mark Mills estimates that the global information communications technologies (ICT) ecosystem uses a total of 1,500 terawatt-hours of power every year, equal to the total electricity generated by Japan and Germany combined.[17] Coal is still the main producer of electricity in the United States, so Mills can say with some justification that "the cloud begins with coal" and that cellphone use contributes to global warming.

We do not see the environmental costs of smartphones and social media. We do not see the global connections of T-shirts or the lights we place on Christmas trees, yet these connections matter. Turning a blind eye to them does not erase the suffering and environmental and social problems that result.

These effects are distant from us. But there are other reasons, more closely tied to the history of consumerism in developed countries, why we should pay attention. Most world religions point to the dangers of too close an attachment to material goods, and myths and stories from Prometheus and Adam and Eve to Shelley's Frankenstein and Wagner's and Tolkein's Rings warn of the dangers of the desire for things and material technologies. But the rise of consumerism from the eighteenth century onward has resulted in societies in which the display of wealth and the emulation of luxury goods has become normalized.[18] The rise of consumerism was initially linked to the international trade in tea, coffee, sugar, tulips, and cotton, and later to the appearance of department stores that marketed items beyond what people really needed. Elites had long competed in the display of prestigious luxury items, but consumerism brought a mass involvement in following and buying the latest fads in clothing, furniture, watches, home decoration, and other goods. Happiness came to be defined in this way despite Protestant strictures in opposite directions, and despite accusations of false superficiality.

The causes of consumerism are complex and, of course, thoroughly entangled.[19] They include increasing wealth, international trade, military dominance and slavery, the influence of the Enlightenment, secularism and the rise of Romanticism, industrialization, and population increase. An important factor was undoubtedly the worry among nineteenth- and twentieth-century capitalists that production would outstrip demand. This made it important for advertising and marketing to stimulate consumers to buy an increasing array of products. Mass marketing focused on making people believe that their identities and well-being depended on being satisfied by the objects of their desires. Human existence came to be tied up with things as objects. Objects had to be disconnected from their production chains and from the inequalities and environmental damage embedded in them—the hard labor and the piles of waste. There are important exceptions, such as the campaigns against purchase of ivory or blood diamonds, but for most things, such as cellphones and Christmas tree lights, slippers and plumbing pipes, we consume without reference to their worldwide consequences. We see objects we desire and grasp for them, shoving into the stores on Black Friday or Boxing Day in a mad frenzy to own and have; in doing so we lose sight of the object as thing. It is in the interest of producers and consumers not to see the thingness of things. It is in their interest to persuade us that only the objects matter.

THE CONNECTIONS OF THINGS

Advertising, social pressure, and our own desires urge us to focus on individual items of consumption separate from their chains of consequences. But is it possible to conceive of and study a thing in itself as object, separate from its connections? Aesthetically, yes: We can, imitating Marcel Duchamp, put a wheel in a museum display and shine lights on it and write labels about it. Is this not the thing in itself as object? Archaeologists and material scientists can analyze the technology used to make a wooden wheel, sample the wood to identify the tree, and provide a dendrochronological date. They can use digital cameras on microscopes to record the wear on this particular wheel, unique in detail from any other wheel. Is this not the study of objects in themselves?

Well, yes, but all these actions actually take place with reference to other things. The museum itself creates a context for our focus on the wheel, and the lighting sets it apart, but the apartness depends on the lighting, the casing, and the label. The label may compare the wheel to other wheels or refer to the wheel's evolution. The scientific analyses require machines, and the research process depends on a vast interconnected web of ideas and theories. The thing depends on a diverse array of other things.

So things connect. I have explained in this chapter how modern capitalism has produced a world in which we are trained to expect things, as objects of our desires, to bring us fulfillment and happiness. We think we know about things and what they do for us. We know we can use things to get us to the moon, build nuclear reactors, and get us to work. We think we are in control of things, but things are often in control of us. We yearn for them even as they draw us ever further into environmental destruction and enormous social inequalities. Things have led us down evolutionary pathways from which we are having difficulty extricating ourselves. Once started, it has become difficult to put out the fire; we have become Prometheus Bound.

Webs of Dependency

I WANT TO MOVE the discussion from networks to entanglements, which means moving from talking about connections to talking about dependency. We have already seen that human relations with things involve dependency. Humans use fire, but that use turned into a dependency that was biological, in the sense that human physiology and even anatomy came to require cooked food. It was also social in that the coming together around hearths selected for sociability. And it was conceptual in that humans developed ideas and myths about fire (such as the Prometheus myth) and about hearths, ovens, furnaces, and so on. Humans use cotton, and one might argue that humans do not really depend on cotton since they could use wool or linen, but the story of cotton told in chapter 4 described vast wealth and global movements, such as the Industrial Revolution, that did indeed depend on cotton and on the machines that produced it. Humans do not really need Christmas tree lights, but the production, consumption, and discarding of the lights employ large numbers of people.

The word "depend" can refer to reliance, and certainly humans rely on things, in the sense of expecting them to be readily available and functional. Humans rely on tools, machines, and buildings. They rely on words and symbols to communicate. But they also depend on them in the sense of being unable to exist without them. Humans cannot be independent of things. Certainly we depend on food and water. We would not last in severe cold or heat without clothes or shelter. We cannot feel without things to feel; we cannot desire without things or people or ideas to desire; we cannot think without words or symbols; we cannot love or hate without things, people, or institutions to love or hate. Human being, as Heidegger demonstrated, is always a being with things.[1] Human thought and perception are always *of* something. We depend on fire and Christmas tree lights to illuminate dark wintry days and nights and to cook Christmas meals. We depend on wheels to get us places, and on cogs to work machines. Our societies and economies could not exist without wheeled vehicles and without machines to build roads and apartment blocks.

The word "depend" has other helpful connotations for a theory of entanglement. The noun "dependency" refers to a territorial unit under the control of another country, often limited and constrained by that dominant country and its interests. In psychological terminology, codependency in a relationship involves one person being supported by another person so that the first person's drug addiction, gambling, or immature or unsocial behavior may continue. In these cases, dependency involves negative notions of constraint or limitation. One person's development is held back by the development of the other. So it is with things. We rely on things like cotton and fire, but we are also constrained by the environmental and social problems that result, whether fuel and resource use or massive inequality. Humans depend on cars, Christmas tree lights, and smartphones but get drawn into negative environmental and social relationships as a result.

Dependency thus involves a double bind. Humans depend on things that depend on humans; humans are both dependent on things and drawn into caring for them. A good example is the domestication of plants, which takes us back to the sickle described in chapter 1. We saw that flint or obsidian sickles were first used for cutting reeds and sedges from the twelfth millennium BCE onward in the Middle East and were used for agricultural harvesting from the ninth millennium BCE. Figures 1.3 and 4.3 showed how the use of sickles became caught up in the operational sequences that produced domesticated plants. The harvesting methods themselves contributed in complex ways to genetic change in wheat and barley. Harvesting and sowing seeds tended to result in the selection of plants with a tough rachis, the central stem that holds the kernels. This selection for a tough rachis meant that the seed heads did not automatically shatter and disperse as they do in the wild state. Greater work was now required by humans in order to obtain the seed from the non-shattering plant. As shown in figure 4.3, this extra work involved threshing and winnowing.[2] From then on, humans were bound into a web of dependencies between humans and things. They depended on harvesting tools and cereals in order to extract sufficient resources from a given unit of land. Things also depended on other things: For example, to function as a sickle, the flint depended on the wood, antler, or horn hafts and on bitumen. These things also depended on humans, as humans repaired and resharpened the sickles, and sourced the flint and obsidian. The domesticated plants, with their nonshattering heads, also now depended on humans since they were less able to reproduce themselves. And humans depended on other humans when they obtained obsidian through networks of exchange, or in the sowing and harvesting of fields and the sharing of the products. Thus the webs of dependency or entanglements spread in all directions: human dependent on human, human dependent on thing, thing dependent on human, and thing dependent on thing—relationships that can be abbreviated as HH, HT, TH, and TT. We can see this more clearly in the example of opium.

Opium

The sad story of opium shows how human dependence on things for positive gain can also produce destructive reliance. The web of positive dependence and negative dependency around opium has brought addiction, wars, imprisonment, crime, and terrorism. Drug dependence may seem too obvious an example of human-thing dependence, but it is a useful case study with which to evaluate the scale and nature of human-thing dependence. Is it going too far to say that we are as dependent on things as some are on drugs? We do not seem able to do without them. Our happiness, identity, and sense of self all depend on things. Can we talk of thing addiction?

Wild poppy variants grow in the Mediterranean and in Anatolia. The domestic poppy is *Papaver somniferum*—although there are also many other varieties. The opium occurs in the seed pod, and its harvesting is exhausting and backbreaking. It has to be done by hand and requires experience and dexterity. The pod is tapped, scored, or lanced to release the sap. The opium is dried into a latex gum and then stored until it is cooked in water, sieved, and dried.

The earliest finds of domesticated poppy are in the European Neolithic, and it is assumed that the seeds were used for food (as poppy seed oil) and the latex to induce sleep.[3] The Sumerians cultivated poppies, and by the seventh century BCE in Assyria, texts refer to opium as a common cure for many ailments. It is also mentioned as a cure in second millennium BCE texts in Egypt. The Greeks used it as a sleeping potion and as a cure, and it was also thought to have spiritual and religious powers. For the Romans it was a cure that also induced sleep, as well as a poison.

The large-scale trade of opium was opened up by Arabic traders in the early medieval period, reaching Iberia, India, and China. Into the sixteenth century CE in Europe, opium was still used for medicine, but there is clearer evidence by then of drug addiction. It was used as an anesthetic during surgery, and through the seventeenth to nineteenth centuries it was mixed with alcohol or wine to make laudanum. In the late eighteenth and early nineteenth centuries, opium greatly interested the Romantic writers, who saw it as producing free passions, imagination, and spontaneity. Coleridge and Elizabeth Barrett Browning used it, and probably Keats, Sir Walter Scott, Byron, Shelley, and Baudelaire as well.

Throughout these eras, the addictions, negative effects, and deaths from overdose were not a great concern, but by 1860 the evolving public health movement in Britain and the more defined and regulated medical profession started to see opium as a harmful drug or poison. There were some early attempts at legislation to restrict availability and usage in the later nineteenth century, but the problem had then been exacerbated by the use of morphine. Chemists seeking to isolate the alkaloids of opium had produced morphine by the early nineteenth century, and it was first marketed commercially in 1827. The invention of the hypodermic syringe in the 1850s—conceived by linking together the idea of a subcutaneous needle with the concept of the syringe—greatly expanded the use of morphine in surgery. But it also led to a great increase in addiction.

In the late nineteenth century, a yet more dangerously addictive opioid drug was discovered. Heroin is not a naturally occurring alkaloid of opium, but it is derived from it. Again it was initially marketed for medical use. But it was easy to produce, only small amounts were needed, and it could be administered with a hypodermic syringe or in tablets. By 1910 its addictiveness was widely recognized, and controls were placed on its availability. Trafficking went underground, and the international illicit trade quickly expanded.

The most egregious and woeful consequences of opium and its derivatives came from their entanglement with imperial expansion into East and South Asia. Opium was grown in China from the first centuries CE, but on a small scale. The colonization of North America had led to the spread of tobacco pipes, and sailors from the Dutch East India Company were the first to put opium in pipes. As Martin Booth writes in his book *Opium: A History*, "Thus was born one of the most evil cultural exchanges in history—opium from the Middle East met the native American Indian pipe."[4] In the early and middle nineteenth century, the British East India Company carefully controlled the trade of opium with China. The imperial Chinese government had banned its importation in the early nineteenth century, but the British East India Company, in collusion with the British government, saw China as a market.

The company controlled the production of opium in India but did not want to sell the opium there, as it had negative effects on its workers. So it decided to focus on export to China despite the Chinese ban on imports. In addition the company wanted access to the tea and silk exports from China but did not want to pay the silver bullion China wanted in exchange. So instead, the company provided cotton and engaged in an illegal opium trade.

The result was immense suffering and addiction. China went to war to stop the trade in the Opium Wars of the 1840s through the 1860s. These wars were an entangled mess involving "the addiction of one empire and the corruption of another,"[5] but they also created important trading centers. "Without opium," Booth writes, "Hong Kong would not have evolved."[6]

Other areas where empire and opium became embroiled from the 1950s onward were Pakistan's North-West Frontier Province and the Golden Triangle area in Laos, Thailand, and Myanmar. Because the US government wanted to halt the spread of communism in Asia, the CIA supported local leaders in the Golden Triangle, including those growing and trading opium. Booth tells us that "the CIA became inextricably entangled with the Golden Triangle opium trade, handling opiate consignments, flying drug runs and tolerantly turning a blind eye to the affairs of their criminal allies."[7] The trade in opium was linked to the trade in arms and gold. The CIA assisted the Hmong (whom they needed to fight against communist infiltration) by flying the drugs in planes of Air America, a CIA-owned carrier. In Vietnam, where the French had gained funding for their fight against the Viet Cong by dealing in opium, the arrival of some five hundred thousand US troops dramatically increased the scale of the trade. The soldiers wanted high-quality heroin, and it was provided by refineries in the Golden Triangle and flown into the war zone by Laotian and Vietnamese military aircraft. In 1971 some 10 to 15 percent of the troops were heroin addicts.

The Golden Triangle remained a major global producer, but a drought in 1978 led to a shift to Pakistan and Afghanistan. In 1979 the CIA started covert operations in Afghanistan to help the resistance against the Soviet occupation, working alongside the Mujaheddin Afghan guerillas. The latter were already involved in opium growing. The CIA's presence is said to have stimulated opium production in this Golden Crescent area, at least in the sense that they turned a blind eye to it. Production and distribution were run by Corsican syndicates that had long been involved in Mafia and "French Connection" opium trading. Between 1962 and 1990, world heroin production tripled. "The global drug trade," Booth writes, "had become more complex and labyrinthine than ever before. There were more players, more sources, more diversity than there had ever been, which meant the problem was all the more difficult to address."[8] In May 1995 the United Nations estimated that 40 to 50 million drug addicts worldwide depended on the heroin produced in the Golden Triangle and Golden Crescent.

Opiates are today the mainstay of narcoterrorism, the use of drug trafficking to further the objectives of terrorist organizations. Drugs are used in the purchase of arms and in the trade in illicit antiquities. The annual volume of drug-trade money laundered globally was $500 billion, $350 billion of which was laundered through the United States. "To eliminate the poppy, massive economic and cultural aid will have to be spent. . . . The cost is astronomical," Booth wrote in 1995.[9] The 1993 US aid package to Colombia to fight drugs was $73 million, and this was just one year for one country. Then there is the difficulty of making sure the money is not lost to corruption. And even if one could eradicate poppy cultivation in one country, another can easily start up. Traffickers make up one of the world's most influential special-interest groups, with an economic power that can cripple smaller countries. In over a dozen countries, drug-generated revenue exceeds government revenue.

Overall, the story of the poppy is again one in which human dependence on things leads to long and gradual realization and manipulation of affordances, and greater entanglements so that more is caught up, ensnared in the chains of conditions and consequences. Humans cannot be blamed for wanting pain relief or ways to calm crying babies. For many farmers the poppy is a source of income. Soldiers who fear death seek relief in heroin. When these dependencies got caught up in the colonial enterprise, the combination yielded a global illegal trade. In the United States, the war on drugs has led to the swelling of the prison economy. Treatment of addicts often involves giving them other opiates or opiatelike drugs to reduce the pain of withdrawal. After morphine was discovered, it was used as a cure for opium addiction; then, when heroin arrived it was said to cure morphinism. This is truly an entrapment, a negative dependency relationship, whether in clinics or prisons. As a species we seem thoroughly trapped in opium and unable to escape.

Addiction can be defined as the compulsive taking of drugs such that a person cannot stop using them without suffering severe symptoms and even death. Opiate dependence is more than a habit. It is a bodily need that is essential to the addict's existence, equivalent to the need for food and water. So in this example, humans depend on opium (HT), but things also depend on other things (TT). Opiate use was greatly increased by the spread of hypodermic syringes and tobacco pipes, and by the whole panoply of medical and drug-use knowledge in which it was embedded, as well as by ships and guns, tea and cotton, which participated in the Chinese opium trade. These thing interdependencies also depended on humans (TH), who domesticated and grew the opium, produced its derivatives, and traded and consumed it. Humans depended on other humans (HH) to grow, produce, trade, and consume, and to exploit and subjugate each other. But always through the manipulation of things, and so we return to HT, the human dependence on things. In all these forms of human-thing dependence, there is both an enabling reliance (whether on income from farming opium or on pain relief) and a disabling dependency that, with opium, can lead to dysfunction and entrapment in a downward spiral of imprisonment and crime.

ENTANGLEMENT

Following from the examples I have given so far, I want to provide a clearer definition of entanglement. Colloquially, the word "entanglement" may have many inflections of meaning. Dictionary definitions usually refer to a mass of threads or strands knotted or coiled together. People talk of entanglements as complicated, unfortunate, obstructing. They are confusing jumbles that are difficult to sort out. Entanglements involve multiple threads making contact at multiple points.

The notion of entanglement shows up in a wide variety of scholarly debates. Biologists often refer to the famous phrase at the end of *Origin of Species* where Darwin writes, "It is interesting to contemplate an entangled bank, clothed with many plants of many kinds, with birds singing on the bushes, with various insects flitting about."[10] Many developments in biology focus on coevolution and symbiosis among humans, plants, and environments. The new Extended Evolutionary Synthesis we saw in chapter 2 envisions culture and genes not as separate forms of transmission but as entering into each other through epigenetic and other processes. Culture and biology are thus thoroughly entangled. But perhaps the field where the word has the most influence today is quantum theory, where "entanglement" describes the mechanical phenomenon in which the quantum states of two or more objects have to be described with reference to each other, even though the individual objects may be spatially separated. This implies that the movement of an object cannot be predicted by reference to local conditions.

In philosophy, the work of Gilles Deleuze and Felix Guattari has long been influential in moving away from hierarchical classification systems and emphasizing the links that flow among entities like rhizomes.[11] Rather than searching for origins and hierarchies of causes, Deleuze and Guattari urge us to focus on the complex ways in which things are caught in webs. Anthropology and the cognitive sciences have focused in recent decades on the ways in which perceptions and notions of self are distributed beyond the physical body. For example, Marilyn Strathern has talked of "enchainment" or distributed personhood. Enchainment, as used by Strathern, refers to Polynesian and Melanesian cultures where persons are "dividuals" or "partible persons"—meaning that they are the products of chains of social acts, so there is no division between the social and individual persona.[12]

In her work on contemporary South Africa, Sarah Nuttall describes the historical entanglements between blacks and whites.[13] The more whites dispossessed blacks, the more the two races depended on each other. In their dependence on blacks, whites erected an ideology of separation and difference—racism. In effect, Nuttall describes a codependency. She rejects simple oppositions between colonizer and colonized, metropole and colony, center and periphery, domination and resistance, and instead talks of an entanglement or "web, carrying with it the notion of interlacing, an intricacy of pattern or circumstance, a membrane that connects." More influential for archaeologists has been the work of Nick Thomas, who writes of Western and non-Western peoples mutually entangled in an array of rights and obligations, people who are "reciprocally dependent" in the exchange of objects in colonial societies. "The notion of entanglement aims to capture the dialectic of international inequalities and local appropriations."[14] Rather than opposing European and indigenous, global and local, capitalist commodity and reciprocal gift, domination and resistance, Thomas has influenced archaeologists to seek a more historically accurate consideration of the complexities of the colonial encounter.[15]

There is also a broader influence from studies of material agency and materiality. Since the 1980s, many archaeologists have followed the work of Pierre Bourdieu on theories of practice and have argued that material culture can be actively employed by people in pursuing their social strategies. The study of material culture and materiality has had a strong impact on the ways archaeologists study humans as agents and things as agents, and a lively debate has occurred on what the notion of things as agents might mean. One of the outcomes of this discussion has been a critique of social theories that place humans at the center without giving a major role to things. Some have proposed a relational archaeology in which identity and causality are seen as dispersed, or have claimed a symmetrical archaeology in which humans and things have equivalent weight.[16]

The calls for a symmetrical archaeology are directly inspired by the work of Bruno Latour. Latour's study of sociomaterial networks has also influenced material entanglement approaches in archaeology, although we will see later that network and entanglement are rather different notions. Sociologists have tended to see the social world as about interpersonal relations. But Latour, John Law, and Karin Knorr-Cetina have come to see how engines, measuring instruments, laboratory probes, and detectors play a part as actors in structuring social relationships. While exploring the production of scientific knowledge in the laboratory, they also argue that similar social/thing processes occur more widely. They focus on the actor networks of big things like the computerized rail transportation system called ARAMIS, but they also look at small things like pipettes, paper blueprints, and computer screens. The aim of this approach, often termed actor network theory (ANT), is to critique apparent fixed and essential dualisms such as truth and falsehood, agency and structure, human and nonhuman, before and after, knowledge and power, context and content, materiality and sociality, activity and passivity.[17] It is not that such divisions don't exist, but that the distinctions are effects or outcomes of the relations between particular human and nonhuman agents. "They are not given in the order of things."[18] Latour is interested in examining what is necessary for something to exist. He looks at things not as self-contained but as needing allies and tributaries.

Another relational approach is provided by the ethnographer Tim Ingold, who looks at the ways in which humans and things coconstruct each other in the practices of making.[19] He explores how things are always changing in relation to humans and their perceptions. Environments are not static places but landscapes that are lived in and moved around; a pebble is different depending on whether it is wet or dry; a basket emerges out of the interactions between reeds and humans. Karen Barad has developed a theory she calls "agential realism," in which objects do not just exist and then interact. Rather they emerge through their "intra-actions." Because everything is so entangled in everything else, when we observe something, we have to make a "cut" that includes and excludes, so that the boundaries of entities are always temporary constructs. In reality there is no interaction between separate entities, only actions within the whole: intra-actions.[20]

Archaeologists have long studied the material properties of things, and the subfield of archaeometry has been devoted to chemical, physical, and biological analysis of how things were made, used, and discarded. They have also devoted much attention to the operational sequences by which things are procured, manufactured, used, and thrown away. While these approaches have been very effective at unpacking the ways in which material things are linked to each other, they have been less effective at linking things to the social world. Things are largely seen as objects of human action rather than as coconstituting the social.[21]

The defining aspect of entanglement with things is that humans depend on things that depend on humans. Put another way, things as we want them have limited ability to reproduce themselves, so in depending on them we become entrapped in their dependence on us. This dependence is amplified because things depend on other things.

Most authors influenced by Latour, actor network theory, and Ingold focus on networks or meshworks of relations.[22] While influenced by these relational approaches, entanglement theory is different because its focus on dependency, including dependencies between material things, leads to a notion of being caught up or entrapped (a cog, a strand, an addict, Prometheus Bound). Rather than simply identifying links in a network, entanglement theory calls for us to pay attention to the complex relations of dependence and entrapment that comprise those links. Another way to define entanglement is to say that it describes the relationship between dependence (often productive and enabling) and dependency (often constraining and limiting). Humans and things, humans and humans, things and things depend on each other, rely on each other, produce each other. But that dependence is in continual tension with boundaries and constraints, as things and humans reach various limits (of resources, of material and social possibility) that are overcome by—that demand—yet further dependence and investment. Entanglement can thus be defined as the dialectic of dependence and dependency between humans and things. Within the networks and flows there is a "caught-up-ness."

The trouble with this definition of entanglement is that it appears to separate human and thing, subject and object in a rather nonrelational way. Human being is always dispersed and distributed into the world, and there is no environment separate from human culture and society. It is difficult to talk of TT dependencies without human involvement, or of HH dependencies without things. Humans, too, are things. Separating out TT, HH, TH, and HT dependencies from overall entanglements is difficult. A more satisfactory definition of entanglement is that it captures the messiness of the flows and counterflows that produce, enchain, and encompass all entities: humans, animals, things, ideas, social institutions. Tables and chairs may seem objectively stable and inert, but they gradually decay and collapse. Cereals and opium have continually changed their form and function. The functioning of the wheel is dispersed through wagons, draft animals, roads, and governments. There are ultimately only flows of matter and energy and information. Entanglement helps us understand these flows by identifying entities and exploring their dependencies. But it is important to recognize that behind the descriptions, the entities themselves are produced out of flows. Opium is always a product of particular circumstances, and it emerges in and out of being by the flows in which it is encompassed.

Your Entangled Car

I want to emphasize two points about these webs of dependency, these entanglements. First, they are heterogeneous. They include an enormous battery of diverse things, from the physical to the metaphysical, from the social to the economic, from inorganic matter to plant and animal. The things that opium brings together include guns, ships, tea, cotton, syringes, pipes, planes, environments, colonialism, empire, governments, and fears of communism. Actor network theory and the new materialisms have trained us to think of the mixed nature of these assemblages, but it is important to emphasize that entanglements are not simply networks but also webs of dependency.

Second, the entanglements are unbounded. Entanglement theory is radically nonreductionist. We saw with opium, for example, that distant events such as the use of pipes in the Americas can suddenly have an impact on its use in China. The invention of the hypodermic syringe can lead to an increase in opiate use in England. These quantum effects can be radical: The Vietnam War increases opium production in the Golden Triangle, and increased fighting in Afghanistan creates a new role for Corsican drug traffickers. Of course, we can attempt to deal with the rise in drug use in the United States by imprisoning users, legalizing drugs, or prescribing more methadone. But none of these approaches has been very successful. From an entanglement perspective, the reason is that these solutions attempt to deal with a small part of the whole. There is no one villain, such as the producers, traffickers, or consumers. The drug problem is diffuse and dispersed, and there is no boundary around it.

The wheel's entanglements are also unbounded. Let us say that you want a new car, but an environmentally friendly one. Cost is not your primary concern, and you are dithering between a hybrid battery/gas vehicle and a purely electric plug-in. At first sight it seems evident that the electric car is more environmentally friendly, since it does not use carbon fuels. But if you follow the chains of entanglements, the picture gets less clear. If your local electricity comes from a coal-fired power plant, your electric car is actually a coal-powered car. Coal-burning power plants emit not just CO_2 but other noxious gases like nitrogen oxides and sulfur dioxide—and in far greater quantities than gas-powered cars. Also, power drains out of an electric battery even when the car is not in use. But a gas-using car depends on refining, processing, and transporting gas, all of which add to the carbon footprint. So the electric car wins out. But we can follow the entanglements still further. Electric cars need to be light, so they use high-performing materials like lithium in their batteries, and they use rare metals throughout—for example, in their ubiquitous magnets. The rare metals come from environmentally destructive mines that use toxic chemicals such as ammonium sulfate. Other mines have high emissions. When all this is taken into account, manufacturing an electric vehicle can produce more carbon emissions than a gas car.

We can pursue still other entanglements. What happens when the car must be broken down and discarded? The lithium-ion batteries in electric cars are so big that a cost-effective method of recycling them does not yet exist as I write this. The situation is continually changing as manufacturing is improved and, for example, people switch to solar power to charge their batteries. And the longer you drive an electric vehicle, the more your zero-emission use (except for the coal you may use to power it) will offset the manufacturing emissions.

So the answer to the question of how green is an electric car is,

"It depends." It depends on where you live, how often you drive it, whether you consider only use or also take manufacture and discard into account. Of course you could follow the entanglements still further. You could examine the costs of every machine and tool used in building a car, and the design process. You could explore the energy used to make the machines used in mines to extract and refine rare earths. You could include all the legacy environmental damage of past gas cars and past coal-fired electricity generators—and so on. The evaluation of things depends on how far one follows the entanglements. Constricting the boundaries gives us a partial view, and only by opening up to consider entanglements in a less bounded way do we reach a more realistic evaluation. The legacy costs of opium and cotton—for example, the wars and suffering they have produced—are tremendous, well beyond the enormous damage they do today.

And yet we cannot escape having a partial view. All studies need to draw boundaries: You cannot study everything. But the entanglement approach encourages us to be as inclusive as possible and also to understand the artificiality of any boundaries we draw. This is particularly important given the myopia with which we twenty-first-century consumers tend to look at things. So many of the consequences of our consumption are hidden from us by distance in space or time, or because they are masked by advertising or buried below pavements and oceans. But it is important to move along the chains of conditions and consequences if we are to understand why human evolution is directional.

CHAPTER 6

The Generation of Change

THE HETEROGENEOUS and unbounded nature of entanglements creates conjunctions, conflicts, and contradictions. There are many examples in the contemporary world. Under President Barack Obama, US relations with China were described as entangled because, on the one hand, the United States and China wanted to reach an agreement on climate change, yet they strongly disagreed on human rights, technology copyright, computer hacking, and other matters. Today there is increasing global starvation even as humans throw away large amounts of food. Attempts to tackle global warming (largely the product of richer Western countries) are in conflict with rising populations (often in poorer countries) and increasing forest clearance. It is a general condition of life that complex phenomena create internal contradictions.

Providing food to a rising global population is a complex process that is rife with conflicting and competing operational chains and the absurdities that result. In 2013 the Institute of Mechanical Engineers in Britain produced a report that estimated that 30 to 50 percent (or 1.2 to 2 billion tons) of all food produced on the planet is lost before reaching a human stomach, and a similar report has been produced for the United States.[1] Why does this occur? Part of the problem occurs early in the supply chain, where inefficient agricultural practices, inadequate infrastructure, limited transportation options, and poor storage capacity lead to squandered harvests and misused land, water, and energy resources. But much of the problem is in marketing and consumption. Supermarkets put lots of food on shelves and end up having to throw much of it away; people buy food cheaply and then discard it. Because of our increasingly pressured lifestyles, supermarkets have invested in ready-to-eat hot food; much of this is not sold but cannot be given to the poor because of health regulations. Overly strict sell-by dates mean that food is often thrown out before its time. The preponderance of buy-one-get-one-free offers causes households to buy more food than they can eat, so some of it spoils. Customer demand for perfect fruits and vegetables causes piles of scratched or misshapen—but still nutritious—produce to end up in the trash. In Britain, some 30 percent of vegetable crops are left unharvested because they do not look good, according to the report by the Institute of Mechanical Engineers. Supermarket chains have sought solutions to combat waste—for example, lobbying the European Union to relax laws against the sale of cosmetically unacceptable produce. There are more innovative solutions as well. The "baby carrot," which has done wonders for carrot sales in the United States, does not pop out of the ground that way but is machine-sculpted from misshapen regular carrots that once would have been thrown out.[2] Overall, however, the problem remains intractable because of the multiple opposing strands and contingent intersecting chains.

I have described many other examples of contradictions and conflicts in entanglements. In chapter 3, the construction of dams came into conflict with concerns about heritage and the environment. In chapter 4, the example of cotton brought together a series of opposites: slavery and free labor, states and markets, colonialism and free trade, industrialization and deindustrialization, plantation and factory, colonizers and colonized. In the mid-nineteenth century in the United States, slavery in the cotton-growing South came into conflict with a rising class of industrialists whose concern for the human condition led them to reject slavery on moral grounds. Chapter 5 described the ongoing conflict between the medical uses of opioids and the harm resulting from addiction.

I have argued that the conjunctions, contradictions, and conflicts in human-thing entanglements derive from their heterogeneous and unbounded nature. Some of these effects are spatial. Distant, apparently unrelated events can set off chain reactions that ripple through the entanglements, as when an artist in Denmark draws a cartoon for a newspaper that causes outbreaks of violence in the Middle East that then affects the price of oil and the numbers of cars on the streets in California. Or Mohamed Bouazizi, a street vendor, sets himself on fire in a market in Tunisia and sets off pro-democracy uprisings throughout the Middle East, again leading to changes in oil prices across the globe. But some of the effects are temporal. For example, the building of dams on the Nile could not avoid bumping into the fact that, five thousand years earlier, Egyptian pharaohs had constructed temples and other monuments along the same river. The legacy of ancient Egypt created complex entanglements that had to be addressed in the present. The same happened with the Yangtze River.

Many of these sources of conflict result from social and economic inequalities, nationalism, colonialism and empire, and religious difference. None of these factors on their own necessarily create conflict. Rather, they create conflict within particular historically specific entanglements. In other words, an overall set of entanglements leads nationalism, for example, to produce exclusion, violence, and conflict. Religious difference leads to war only within particular historically derived circumstances. Class differences and inequalities in the social relations of production do not by themselves generate change. They do so within particular conjunctions. We must consider the specific set of entanglements rather than assume an inherent generative character for particular social forms.

Similarly, entanglements are experienced as more entrapping or less so depending on context. Often the ways in which they are lived are backward-looking and historically derived. For example, earlier British and French colonial entrapments in the Middle East and struggles to control oil from the early 1900s onward contributed to the trauma of September 11, 2001, and thus to the focus on security that dominates so much of our lives and has contributed to right-wing populist movements and to anti-immigration politics. Capitalism in the twentieth and twenty-first centuries depends on mechanization that has caused many people on assembly lines to be replaced with robots, with the result that many people and communities suffer the effects of joblessness, fueling a political move to the right—restricting immigration and encouraging the building of a border wall. Or the very negative experiences of fascist Italy and Germany in the 1930s and 1940s led Central European academics to the goal of a nonpolitical empiricism from the 1950s to the 1980s. Or the failures of the Soviet Union led to the revolutions of 1989 and the move toward democracy and liberal market economies in Eastern European countries. In all these cases, past entanglements influenced the direction of later political change.

The historical situation of entanglements includes the perturbations caused by booms and busts. In chapter 1 I described Ian Morris's work demonstrating long-term exponential increases in human energy capture, but I also pointed to other evidence for shorter-term variations. There is much evidence of "booms and busts" in prehistoric Europe, with the "boom" areas shifting around over time. In Stephen Shennan's European examples, the booms and busts are identified as cycles of population increase and decrease.[3] But we have seen several other types of changing fortunes as well. For example, at the end of the Pleistocene in the Middle East, the climatic warming that ended the last Ice Age was broken by a short colder and drier spell called the Younger Dryas. Early Natufian societies that had grown settled during the first warm phase became mobile again during the Younger Dryas. Yet they did not fully return to their earlier mobility. Too much had by now been caught up in a sense of place, and they retained special relationships with older settled sites through burial practices. When warmer conditions returned after the Younger Dryas, the Natufians quickly reestablished their more sedentary communities. Climatic perturbations create problems that have to be solved, but societies never return exactly to earlier states. In the examples I have shown from more recent centuries, we have seen perturbations caused by cycles in rising and falling prices, the availability of goods, or the availability of labor. These ups and downs unsettle entanglements, add to the messiness, and create contradictions that need resolving.

REPEATING PATTERNS OF CHANGE IN HUMAN-THING ENTANGLEMENTS

All this messiness and uncertainty might appear to lead simply to more contingency and uncertainty, with no pattern in the flow of change. But I want to argue that there is a general pattern: The messiness and contradictions draw humans into ever greater human-thing entanglements. The examples I have provided in this book show a repeated pattern of human dependence on things that depend on other things. These things are themselves caught up in entanglements, whose contradictions and conjunctions draw humans into greater dependence on things. It is a generative spiral.

Take, for example, the long-term human dependence on cereals. In the later Pleistocene, from around 22,000 BCE onward, humans increasingly used grinding stones to obtain more nutrients from plants.[4] The use of these stones, which were difficult to move, came into conjunction with other food processing strategies. For example, many food plants need roasting or boiling, so a hearth, oven, or fire is needed along with the grinding stones. The hearths and grinding stones could be used for multiple purposes, and the conflicts between these different purposes could be resolved by bringing the plants to the hearths and grinding stones rather than the other way round. It made sense to designate a central spot for all the processes involved in intensive food production. But this solution came into conflict with a mobile lifestyle. Rather than maintaining mobility and dispersing different processing functions and their tools in various locations, humans built more stable houses and settlements that could function as central places for intensive plant food processing. But this meant the gradual shift to sedentary life. So here humans depend on things such as grinding stones (Humans depend on Things—HT) that become entangled with other things such as hearths and tools (Things depend on Things—TT). These Things then depend on Humans (TH) to be managed in conjunction with each other. To solve this management problem, humans get drawn into building more stable houses and sedentary societies (Humans depend on Things—HT). HT leads to TT and TH which leads to greater HT; or yet more simply HT → (TT and TH) → greater HT.

Within this more intensive food processing, wild cereals were attractive because they provided a greater return for the investment of energy and because they were naturally well packaged in husks for storage. Humans invested in processing cereals, introduced a number of new technologies, and transferred the use of sickles from reeds and sedges to the harvesting of cereals. These changes became entangled with the biology of the plants to favor cereals with a tough rachis. These new domesticated forms could not reproduce by themselves and they needed harder work to thresh and winnow. So humans got drawn into a long-term spiral of increasing investment in cereals that produced greater need for labor-saving devices. Thus humans depend on things (wild cereals) that depend on other things (sickles) and are entangled with human behavior in a way that the cereals are domesticated and come to depend on humans (to thresh, winnow and so on) leading to greater human dependence on cereals. So once again HT → (TT and TH) → greater HT.

The same can be said of more recent harvesting methods. In his novel *Tess of the d'Urbervilles*, Thomas Hardy describes how the introduction of a steam thresher affects a Wessex farm worked by Tess and other women:

> Close under the eaves of the stack, and as yet barely visible was the red tyrant that the women had come to serve—a timber-framed construction, with straps and wheels appertaining—the threshing-machine, which, whilst it was going, kept up a despotic demand upon the endurance of their muscles and nerves.

> A little way off there was another indistinct figure; this
> one black, with a sustained hiss that spoke of strength
> very much in reserve. The long chimney running up
> beside an ash-tree, and the warmth which radiated from
> the spot, explained . . . that here was the engine which
> was to act as the *primum mobile* of this little world.[5]

Here the machine is a tyrant that destroys the lives of farm-
workers, objectifies and enslaves them.[6] The machine was created
to decrease the labor of the harvest but it actually increased it, so
that the rural people regretted the passing of the old days "when
everything, even to the winnowing, was effected by hand-labour."[7]
As Zena Meadowsong writes, "Far more brutal and exhausting
than the laborious methods it was built to replace, the machine
is inhuman in its strength and dehumanizing to those who tend
it." Tess has to stand on top of the machine and pass the sheaves
to the engine man. While others can occasionally rest, Hardy
describes how Tess has no break from the machine's demands.
"As the drum [of the thresher] never stopped, the man who fed it
could not stop, and she, who had to supply the man with untied
sheaves, could not stop either." The machine's hum "increased to
a raving whenever the supply of corn fell short of the regular quantity." When the humming and whirling of the machine finally
finishes, Tess's "knees tremble so wretchedly with the shaking of
the machine that she [can] scarcely walk" away from it.

Modern readers may find this account overwritten, but it captures the reactions to industrialization that also caused people to
attack spinning machines. Steam threshers and other mechaniza-
tions of ploughing, reaping, and winnowing in the eighteenth and
nineteenth centuries were the results of the shift from enclosed to
open fields (allowing larger fields in which machines could work);

of the Napoleonic and American Civil Wars; of the need to produce more at lower cost as a result of competitive international
trade; of increasing urban populations, a process that was itself
linked to the mechanization of cotton spinning; of the shortage
of rural labor in the mid 19th century;[8] and of the high price of
horses. In the past, threshing had been done with flails; it was
a slow process lasting much of the winter, but it provided rural
employment. So humans depended on things (threshing flails),
but these were entangled in a contradictory way with land organization, colonialism, war, and the rise of industrial cities. A
solution had to be found that enabled farmers to provide larger
amounts of grain at a profit. So humans were drawn in to greater
dependence on things (threshing machines). The farm workers
who were forced out of their jobs blamed the machines, and the
"feeling reached its climax in the winter of 1830, when bands
of starving farm labourers roamed the countryside smashing
threshing-machines and burning hay-ricks."[9]

Hardy, a consistent critic of mechanization, captures well the
ways in which humans get trapped into harder work and changed
lifestyles through their reliance on machines to lower costs and
increase production. The descendants of Hardy's steam threshers
are the massive combine harvesters seen on farms today. These
farms often seem deserted. The laborers have been replaced by
faster and more efficient machines that allow more to be obtained
from a unit of land at lower cost. But the machines entrap in new
ways, requiring large debt (the biggest new ones cost as much as
a McMansion), and locking farmers into the global supply chains
of large multinational companies. From mechanized feedlots to
automatic irrigation systems to agricultural machinery, indus-
trialized agriculture places ever-greater demands on fossil fuel,
water, and topsoil resources. All this mechanized equipment

requires petroleum, as do synthetic pesticides and fertilizers, increasingly tying the cost of growing food to the price of oil. Ground water levels have been disrupted, and the erosion of topsoil has led to a dependence on nitrogen supplements. These entanglements have harmful impacts on the environment. As noted in chapter 1, agriculture is responsible for 18 percent of the total release of greenhouse gases worldwide.

The despot in all this is not the machine but the entanglements. The entrapment occurs because humans depend on machines to increase profits that are being undercut by global trade or undermined by lack of labor or other changes occurring in the entanglements. But the new agricultural innovations create further entanglements, which again require new technologies. Humans get drawn deeper and deeper into the lives of things. Whether we are talking about the Neolithic or modern capitalism, there is a continual spiral in which humans depend on things in a way that draws them into greater dependence on things.

We have seen many examples of this process throughout this book. I want to spell out some of the examples so that the generality of the process can be appreciated. In the instance of the Aswan High Dam, there were interactions with heritage sites. Humans over millennia have often destroyed evidence of their past activities; they are perfectly capable of destroying heritage sites. But over recent centuries many humans in developed countries have come to the view that heritage needs conserving. By the time of the Aswan High Dam, humans had become so entangled in heritage that they got drawn into lifting and relocating massive monuments. So humans depended on a thing (dam) that was entangled with other things (heritage sites) that depended on humans if they were to be preserved. In resolving the contradictions between generating electricity and preserving monuments, humans got drawn in to greater human dependence on things such as the mechanisms to lift monuments and the bureaucracies of the UNESCO World Heritage Center (HT → (TT and TH) → greater HT).

In the example of cotton discussed in chapter 4, in the period leading up to 1780, relatively high wages were paid to workers in England, and these were contradicted or undermined by the low wages paid in India such that the English manufacturers had difficulty competing with the Indian output. Spinning in England had been done using a spinning jenny in the home. But the loss of profits caused by competition with India stimulated the invention of the water frame and the ensuing mules. The new technologies changed the equation by mechanizing and speeding up production. Cotton could now be produced cheaply and on an increasingly massive scale. So here human dependence on a thing (cotton) was dependent on another thing (spinning jenny) that was entangled with a mode of production that was contradicted by a different mode in faraway India. As a result, new spinning technologies had to be invented (spinning depended on human intervention), and so humans were drawn into greater human dependence on things (cotton produced by mules using water, coal, and steam power). There were of course further consequences, such as drawing humans into cities and into health problems (HT → (TT and TH) → greater HT).

In the case of the wheel and the car described throughout this book, humans depend on a thing (car), which is entangled with another thing (the environment), leading to global warming. As a result, cars and the environment depend on humans to create inventions such as hybrid and electric cars and renewable sources of energy. Humans thus become dependent on new things and new entanglements (HT → (TT and TH) → greater HT).

A CUMULATIVE PROCESS WAITS FOR
THINGS TO HAPPEN

One way of conceiving of entanglements is as a web of dependencies (as in figure 6.1). But there is a danger that this conception loses the temporality inherent in entanglements. A key idea is that the various steps in operational sequences have to "wait for" each other. Thus, another way of conceiving of entanglements is as a confluence: a set of flows merging into each other. For example, a human puts energy into a stone to create a rough, flat surface that the human can use for the work of grinding plants. By doing so, the human becomes caught up in looking after and replacing the grinding stone. And each step along the way has to "wait for" the previous step in a cumulative process. There is a temporal flow of energy from sourcing to making and using and discarding. These flows get caught up in other flows. In the Epi-Paleolithic in the Middle East, grinding stones got tied up with plant processing, cooking, hearths, and houses. The processing of a particular plant had to "wait for" the grinding stone to be made, the hearth to be lit. The different operational sequences or flows were caught up in each other. figure 6.2 provides an example of entanglement conceived in this way.

I have put "waiting for" in quotes because the term anthropomorphizes the operational flows or sequences. The different parts of the flows do not actually wait for each other, but the term is useful in that it shows how entanglements are very time-dependent. This is true for operational sequences but also at a larger scale. The production of cotton got entangled with flows of imperial rule and cascades of humans being shipped across the Atlantic. The processing of opium and the circulation of narcotics got caught up at various times with the spread of tobacco

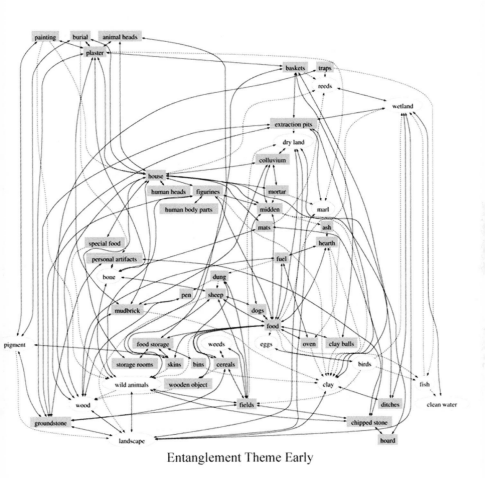

Entanglement Theme Early

Figure 6.1. Formal network and tanglegram (overleaf) for clay use at Çatalhöyük. Source: Author, reprinted from Hodder and Mol, 3.

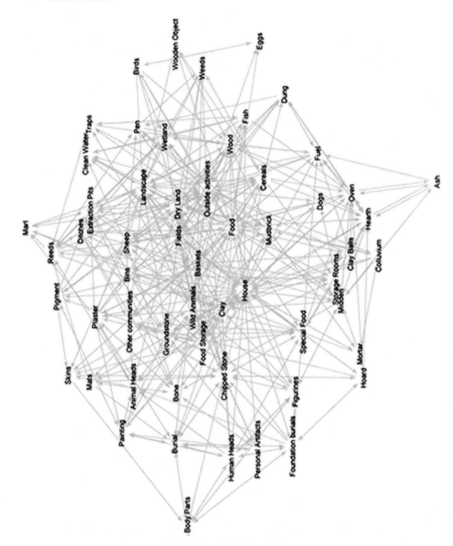

Figure 6.1. Tanglegram for clay use at Çatalhöyük.
Source: Author, reprinted from Hodder and Mol, 3.

Figure 6.2. (facing page) Operational chains for cooking with clay balls and
their cross-connections. Source: Author, reprinted from Hodder and Mol, 6.

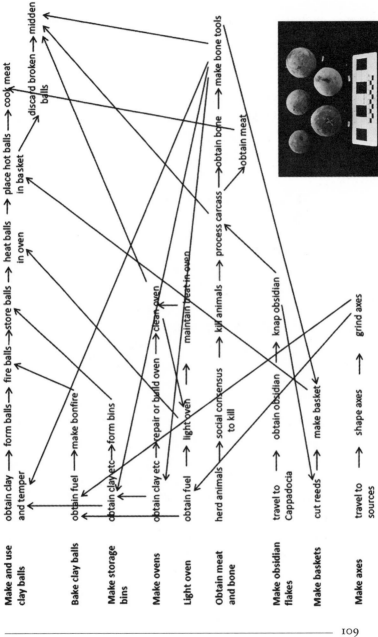

pipe smoking, the Western fear of communism, and the spread of international terrorism. The construction of the Aswan High Dam brought flows of water and electricity into conjunction with flows of information about heritage.

There are thus intertwined flows of energy, power, and information, as well as flows of disease, desire, or spirituality. Entanglements, the lacing together of these flows, lead to odd conjunctions that generate change.

ENTANGLEMENTS PRODUCE DIFFERENT HUMAN GOALS

A clear pattern emerges from the examples described in this book. Humans depend on things that are entangled in consequences (chains of conjunctions, often contradictory and conflicting) that draw humans into greater dependence on things. This increase in human dependence on things occurs because of the heterogeneity and unbounded nature of entanglements, the unruliness of things, and the historical perturbations to which they are subject.

In chapter 3, comparing the dam-building activities of beavers and humans, I argued that the most important difference was that humans follow further down the chains of consequences in order to fix things that go wrong. Humans are both *sapiens* and *faber*, and this dual character is what draws them into ever greater entanglements. Over time, the chains of dependencies between humans and things as well as between things and things have become global and highly complex. Things bring together and assemble numerous different types of processes, resulting in messy, tangled sets of dependencies. As they get drawn down these far-flung and complex intersecting operational chains, humans continually have to deal with complex uncertainty. Things are always falling apart. Dams keep breaking, cotton needs more labor, opium leads to addiction, cars contribute to global warming. We are subject to both dependence and dependency: We depend on things but are constrained by the problems of things and their demands. There is a dialectic relationship between dependence and dependency that leads to change.

Returning to the arguments presented in chapter 2, might we say that humans' increased use of things is simply the result of human goals and intentions? Surely humans just want to better their lot, and they take the opportunity to do so whenever they can? As a result there is a gradual increase in use of and dependence on things, and more stuff. The trouble with this claim is that it begs the question. It is teleological. It assumes what it should be setting out to prove: Why do humans have the goal of material progress? Many societies past and present, and many recent religious traditions, have denounced the amassing of material wealth. Our contemporary drive to be material consumers and to seek technological solutions is firmly embedded within mercantile and industrial capitalism. So it seems we have to understand human goals and intentions as occurring within particular entanglements.

A final point: If the human dependence on things is entangled with thing dependence on things and human and thing dependence on humans along chains of dependency and consequence, then for every change in the human dependence on things, there must be multiple changes along the flows and chains. If humans depend on a thing, and that thing is linked to many other things, then any change in the human dependence on things will get multiplied along the strands and threads of entanglements. As a result, change in entanglements as a whole will always be exponential. We can finally return to the question of why human-thing entanglements always increase, and why over time there is more stuff and more human dependence on stuff.

Path Dependence and Two Forms of Directionality

To sleep, perchance to Dream; Aye, there's the rub,
For in that sleep of death, what dreams may come,
When we have shuffled off this mortal coil,
Must give us pause.
—*HAMLET*, ACT 3, SCENE I

IN CHAPTER 6 I argued that human dependence on things (HT) leads to thing dependence on other things (TT) and thing dependence on humans (TH), producing greater human dependence on things (HT). Human-thing relations generate change. Put another way, humans use things to solve problems and rely on things to get things done, but they get drawn into the consequences of things and their webs of interdependence. These consequences produce contradictions, conflicts, and contingencies that lead to problems that humans again deal with by using things. Thus HT → (TT and TH) → greater HT.

There is thus a continual dialectical drive toward change. But rates of change vary depending on the scale of the human-thing dependencies. Early humans used very simple things with limited chains of consequences. Making stone or wooden tools did not involve humans in extensive operational sequences and large amounts of equipment. Few problems resulted, and humans were not drawn into further dependence on things. But slowly at first and then gradually faster and faster, they acquired more and more things in an expanding cumulative cone (figure 7.1). Every new dependence brought yet further chains of consequences. The result was compound, and the rate of change exponential.

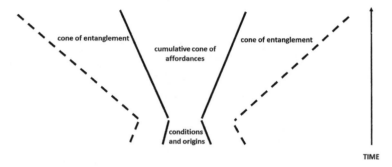

Figure 7.1. Through time there is a gradually increasing exploitation of the affordances of things, associated with an expanding cone of entanglements. Source: Author.

The increasing exploitation of the affordances of things also increases the numbers of entanglements of those things. Thus, as spinning machines became elaborated and more efficient, their entanglements multiplied. As the wheel became diversified, specialized, and differentiated, each new type or function of wheels had its own entanglement. The more types of wheel, the more entanglements. The entanglements thus multiplied exponentially. The cone of increasing elaboration of a thing is associated with a cone of increasing elaboration of entanglement, creating exponential growth (figure 7.2).

We can, of course, disentangle ourselves from things. We can

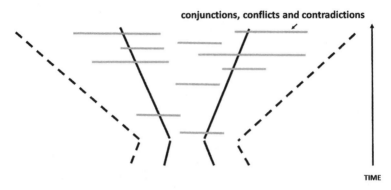

Figure 7.2. Through time, conjunctions, conflicts and contradictions force change and response, pulling entanglements in ever-expanding directions. Source: Author.

withdraw and lead simpler, cutoff lives. We escape into Shakespeare's sleep of dreams. The English language offers remarkably many phrases with which to describe shaking off the coils that bind us. Some are static and spatial, like "thinking outside the box," but many imply a change in direction, an escape from a flow in which one is caught: take the path less trodden, off-piste, off the beaten track, take a fresh tack, off the tracks, swimming upstream, buck the trend, go against the flow or current, and against the grain.

During the Younger Dryas in the Middle East at the end of the Pleistocene, Natufian communities partially disentangled themselves from sedentary villages and returned to a more mobile way of life, but only for a time. As warmer climes returned, sedentary life was again taken up. In Scandinavia, after the adoption of farming, some societies returned for a time to a more hunting-and-gathering mode of existence. In the modern world, we can disentangle ourselves by starting new communities with utopian ideals, forming ascetic cults, meditating, and turning away from the world. We can break machines and reject new technologies. In the same way that fire from wood or coal is increasingly being replaced by other forms of heating and energy, some people are escaping from dependence on cereals and agriculture by embracing a Paleo diet. There are movements such as new consumerism, ethical shopping, and fair trade. Many people manage without cars, and we would survive without cotton T-shirts; indeed, synthetic fabrics have increasingly replaced cotton apparel. Most people never become addicted to opiates and opioids, and some who do manage to rehabilitate.

But it is not easy. Breaking a heroin addiction involves breaking with other addicts, providers, and the webs of debt, violence, and exploitation in which the heroin is entangled. We could stop wearing cotton T-shirts, but a massively profitable global industry would lobby and cajole. While some people can manage without cars, the entanglements of cars are too great to allow an easy exit. It is not just the manufacturers and oil companies, but also the whole social and economic system that depends on cars and other wheeled vehicles. Many of us live in areas where the built environment—widely separated houses, widely dispersed working and shopping areas with stores and offices separated by vast parking lots—has been created specifically for people in cars. We may also depend on cars because we cannot afford to live near where we work and so have to commute, in regions where there is insufficient public transport. We could try to walk and ride buses everywhere, but such solutions are often not possible. We could all try and live on preagricultural diets of meat and berries and greens, but the economics do not add up: It would be impossible to feed the current global population (of humans and cattle). In all these cases, the entanglements of things and the amount invested in them make "going back" and disentangling very difficult.

So, too, with poverty and inequality. At the start of this book I noted that the long-term increase in human-thing entanglement and the amount of stuff humans have accumulated were associated with massive inequality. Today income inequality is increasing in the United States, while social mobility is decreasing.[1] In the United States, there is a strong ideology that individuals who grow up poor can grow rich through hard work. But all societies, including American society, show only slight amounts of upward mobility. Why is this? Why can't humans disentangle themselves from poverty?

Charles Gore defines the poverty trap as "a situation in which poverty has effects which act as causes of poverty. There are thus vicious circles, processes of circular and cumulative causation, in which poverty outcomes reinforce themselves."[2] He argues that poverty traps exist at different scales, from the household and community to the national and global, and that they are caused by a number of compounding factors. Alleviating poverty involves more than simply increasing income. Aid must also address entanglements with other processes, such as environmental degradation, poor health, war and insecurity, corruption, lack of infrastructure, poor education, and lack of skills. Once a path has been taken that leads to greater inequality, all the entanglements and investments that have been entered into make it very difficult to go back. Aid or low-income housing may assist those in poverty, but the problem is too large and multifaceted for simple solutions to be effective. There is also the need to invest in education so that people can get and retain higher-skilled and higher-income jobs, in better health so that they can pursue education and work, and so on throughout the entrapping entanglements. This notion of getting stuck down pathways is captured by the phrase "you can't disinvent the wheel" and by the notion of path dependency.

PATH DEPENDENCY

I have long been skeptical when people say that an inability to change or innovate is due to "culture" or "tradition"—even more so when they say that it results from some "innate" characteristic of a person or group. I have always been struck by how readily humans can adapt to cultural difference. Doing research in East Africa I often found people carrying around different sets of bodily adornment so that they could change affiliation as they walked from compound to compound across an ethnically divided landscape.[3] In this book I have argued that it is not culture, tradition, or character that holds things together, but a lattice of dependencies. Certainly culture, tradition, and character get caught up in these entanglements and contribute to them, but only insofar as they operate within the chains and channels of human-thing dependencies.

We are all aware of a biological directionality to our lives. As we grow older we have to face the fact that we can no longer have children or climb a mountain. The biological clock never runs in reverse. But on top of that, we make life choices that create pathways that are difficult to change. For example, if we gain a doctorate in classical archaeology, it is then very difficult to retrain as a medical doctor; training in the arts and sciences is very different, loans have been taken out and have to be repaid, getting into medical schools is difficult, and the training is long. The entanglements of having taken one path make changing paths difficult. Entanglements are made up of biological, social, material, and many other strands. Together these dependencies mean that to change one thing involves changing many other things and processes. Entanglements are so far-reaching and nonsystemic that disentangling is difficult, so change tends to build on what is already there rather than start over. Change is thus cumulative: We saw this, for instance, in the gradual development of spinning machines. The design of the latter was initially based on hand-spinning techniques, and on flax-spinning machines. The rise of cotton manufacturing in Britain was built on Britain's existing role in sea trade and contacts with India. The development of spinning machines depended on the history of waterwheels, steam engines, and later, electricity. The timeline for the invention of the automobile was similarly cumulative. The invention of the wheel in the fourth millennium BCE was itself dependent on the earlier domestication of cattle, which was linked to the domestication of plants. The modern car thus owes a deep debt to these prehistoric events. But it is more immediately a result of the sketches of a self-propelled car by Leonardo da Vinci, the first tractor in France in 1769, Brown's internal combustion engine in 1823, Lenoir's 1863 "horseless carriage," Otto's 1867 use of a piston chamber, Benz's gas-engine automobile, the introduction of a steering wheel to replace a steering tiller in 1900, the telescope shock absorber in 1901, drum brakes in 1902, the use of the assembly line by Ford to produce cheaper cars, the electric starter in 1911, the car radio in 1924, the power steering system in 1926, flashing turn signals in 1935, the Interstate Highway Act in 1956 that created a network of highways across the United States, the seat-belt laws that started to be introduced in the United States in 1962, the airbags that came into operation in 1974, the GPS introduced in cars in 1996. The list could be extended without end. Humans over millennia (at least in some parts of the world) embarked on a pathway of increased dependence on wheeled transport. In the nineteenth and twentieth centuries, they took a pathway very much dependent on gas-powered cars. Electric cars were invented in the later nineteenth century, but they have only come into their own in the twenty-first century. For over a hundred years, humans gradually developed and accumulated materials, skills, knowledge, road systems, and trade that allowed a particular pathway to be developed. Today it is proving difficult to turn back to a full dependence on nongas cars.

Versions of the idea that it is difficult to switch pathways or to go back once a particular pathway has been taken occur in a number of disciplines, including the biological study of cell development, political science, and sociology.[4] Sociologists depict switching paths as difficult because of linked infrastructure, law and convention, systems of authority, and power differentials. In economics, Paul David studied the persistence of the QWERTY arrangement of letters on a typewriter keyboard and showed that ways of training typists had become so institutionalized (for example, in secretarial typing schools and manuals) that switching to another arrangement would have been almost impossible.[5] From an economist's point of view, the costs of switching to a new path increase over time, and so do the costs of exit from a particular strategy, creating a lock-in effect. Brian Arthur argues that the idea of path dependency is related to the idea of increasing returns: Once a firm gets a leading market share through small historical accidents, the lead can create a positive feedback process so that that the firm outcompetes others.[6] For example, Packard and

Hewlett happened to start their high-tech company in Palo Alto, California, and because they were successful, it became advantageous for other tech firms to locate nearby. There was a runaway effect that ultimately caused the region south of San Francisco to be nicknamed Silicon Valley. Mutually reinforcing mechanisms mean that a small initial advantage can multiply and create benefits. From the point of view of entanglement theory, however, these accounts of path dependency pay too little attention to all the linked consequences of a particular choice of action. They concentrate on specific domains (economics, sociology, politics) rather than on the full range of linkages, and they do not closely define the linkages themselves except in cost-benefit terms: They do not explore complex historical dependencies in the ways that I have sketched for cotton. In addition, many of these models treat the path initially chosen as "random," or at least very much affected by chance. This seems unrealistic. In fact, the path initially chosen is itself entangled.

A good example of how cumulative entanglements lead to path dependency can be seen in European towns. Many have road systems that derive from Roman road systems established two thousand years ago, which in turn often followed pre-Roman paths and trackways. The well-built Roman roads attracted settlement along their course, and they remained thoroughfares as buildings were built and rebuilt around them. The thoroughfare often remained long after any trace of the Roman road had been dug away. This is as true in Rome itself as in London or Cambridge or Aix en Provence. In all these cases, the road alignment was entangled with access to buildings, so that changing the route would have meant changing the buildings along the route. This was obviously difficult and expensive. It was easier to build new buildings along the existing road alignment. Of course rebuilding all the buildings was possible for a sufficiently strong centralized power, like Napoleon III when he and Haussman remade Paris in the mid-nineteenth century.

In chapter 3 I argued that recent work in biology suggests that genotypes impose constraints on evolution, and many of these constraints are legacies of evolutionary paths taken long ago. All species have structural forms that limit the range of variation. This is a similar argument to the notion that ancient road systems can have a legacy effect on the way towns develop. In the Neolithic "town" of Çatalhöyük, buildings were so tightly packed together that each time a new one was constructed, it had to work within the space shaped by the surrounding buildings. More generally, the locations of buildings and monuments affect movement and access across landscapes. Industrial zones attract industries. Research universities attract high-tech businesses to grow up around them. One of the reasons I chose the examples used in this book is that they involve pathways taken many thousands of years ago that still have legacy effects—the wheel, fire, cotton, opium, harvesting tools, the idea of progress.

But for all these examples, the early stages of use were very slow in developing. Affordances were only gradually realized. Real lock-in can take a long time. Although the wheel was in use by the fourth millennium BCE in Eurasia, the real takeoff in dependence on wheels for transport and machinery was in the nineteenth century CE. In chapter 4 I described the progress of cotton from a slow start to a massive contemporary global industry. It took a very long time for the conditions to be met that allowed the full affordances of cotton to be realized. Opium offered a similar story. Poppy seeds were domesticated in the European Neolithic, and a large-scale trade in opium existed by the medieval period. But it was the entangling of opium and its derivatives with imperial expansion into East and South Asia that led to the widest entanglements.

It often takes time for the affordances of things to be realized. It took time for opium to be transformed into morphine and then heroin, or for different types of harvesting tools to be brought together in one large machine, the combine harvester. The idea of progress was present in the classical world, but it only came into widespread use in conjunction with imperialism and capitalism, especially in the eighteenth and nineteenth centuries CE. Notions of progress declined in popularity in much of the twentieth century. Certainly, within entanglements, things can decrease in importance and centrality. For many millennia the fate of the wheel and the wheeled vehicle were tied to the horse; only with the rise of steam power and the internal combustion engine did the numbers of horses kept on the streets of cities and towns decline.

Archaeologists like to identify phases in the development of cultures and civilizations. They often refer to Early, Middle, and Late phases, or to Archaic, Formative, Pre-Classic, Classic, and Post-Classic stages. They describe the "battleship" shape of the curves in the frequencies of traits as they are introduced, become common, and decline. Increases occur as the affordances of things are gradually realized, and decreases occur as entanglements change and the thing loses its role. The rise and fall are entirely dependent on the entanglements. There is no inherent adoption-to-rejection process. All depends on the particular thing and its affordances, and on the surrounding entanglements. Cotton and poppies, for instance, are both difficult crops to harvest and process, and their use through time depended on the gradual accumulation of new techniques within new imperial strategies. The interaction between the affordances of things and the entanglements within which they are realized leads to particular pathways being taken at particular historical moments. Once takeoff occurs, there is more lock-in, and greater difficulty in turning back, because so much has become entangled.

CHANGE IS DIRECTIONAL IN TWO SENSES

There are two forms of directionality in human-thing entanglements.

Specific Directionality

Specific directionality results from the particular things that are caught up in webs of dependency and on the cumulative nature of development. For example, it is often argued that the wheel developed in Eurasia because there were draft animals to pull carts and wagons. Americans in pre-Columbian times did not use wheeled vehicles for transport. This is not because they could not come up with the idea of the wheel, as the existence of toy wheeled vehicles in Olmec sites in southern Mexico makes clear.[7] But the lack of effective draft animals in the Americas, along with other entangled factors, took the Americas and Eurasia on very different pathways. Wheels, draft animals, and their associated diseases, along with the development of steel and firearms, all contributed to a specific pathway in Eurasia that differed from that of the Americas. European ideas of progress contributed to the treatment of societies in the Americas, and after colonization the exploitation of sugar and cotton led to the introduction of slavery. This, too, created a distinct pathway.

As another example, in chapter 6 I showed how, well before the agricultural transformation in the Middle East, grinding stones played a part in entrapping humans into greater entanglements in the ways food was processed. Grinding stones allowed humans to turn grain into flour, which could be mixed with water, milk, or other ingredients to make breads and cakes. In the Middle East the main domesticates (wheat, barley, and rye) all contained the protein gluten and thus could be made into leavened breads. An entanglement between grinding stones and cereals containing gluten led to a bread-based foodway that continues today. In East Asia, on the other hand, there was a very early focus on boiling and steaming. These contrasting food-processing technologies preceded agriculture but became more elaborate as agriculture increased. As Dorian Fuller and Mike Rowlands write, "These traditions also have very different approaches to the supernatural, with a western emphasis on sacrificial smoke feeding distant gods and ritual food sharing promoting community solidarity, and an eastern emphasis on ancestral spirits kept close to the living through the commensal sharing of foods; this has promoted 'sticky foods' including the evolution of glutinous rice and millets."[8]

Archaeologists and historians have studied countless local and regional versions of such differences. Building culture history is one of our main tasks. But there is a danger that we reify "culture" so that it becomes itself a thing, rather than seeing things as entangled in a way that creates pathways. Rather than objectifying culture, we need to look at the chains of consequences that have led to particular responses, and at the cumulative entrapment of humans and things in particular ways of doing things. We have seen how particular circumstances led Britain to become an industrial powerhouse in the late eighteenth and nineteenth centuries; how soil, climate, and particular imperial strategies led to opium production as a cultural trait of the Golden Triangle and the Golden Crescent in Southeast Asia and Afghanistan; and how an international cotton network produced slave "cultures" whose impact in the Americas is still a major social and political force.

General Directionality

As I showed at the start of this book, it is difficult to refute the archaeological evidence for an exponential increase through time in humans' ability to capture energy and in the amount of material things and the entangled webs of dependency. Whatever the specific directionalities that are taken, there remains an overall trend: an increase in the amount of human-made material stuff and in the scale of entanglement. These trends are the logical result of the principles I have described in this book. The overall directionality results from the human dependence on things and from the "thingness of things." The distinctive component of my approach is that it gives a central role to things and to the ways in which things pull other things and humans toward them. Things assemble a diversity of other things or processes. Humans get caught up in managing the entanglements on which they depend. But in doing so, they become bound to something unbounded. Entanglements are heterogeneous and open-ended, incorporating many different types of biological, social, material, and ideological processes. There are thus always contradictions in the processes that things draw together. Entanglements always lead to conflicts, problems, and contingent interactions.

Humans are caught up in things, humans and things are dependent on each other, and so humans have to deal with consequences and with the quantum events that seem to come "out of left field," difficult to predict and control. In fixing problems and in reacting to events, humans do what they have always done: Find another thing, tweak the machine, and manage with what resources are available. As Timothy Mitchell puts it, "Technical change does not remove uncertainties, as the conventional view of science proposes—it causes them to proliferate."[9] In chapter 6 I described the resulting generative spiral with the notation HT → (TT and TH) → greater HT. The entanglements expand and the human entrapment in things grows. At some point, when the entanglements around any one thing (such as a wheel) become too large, it becomes impossible to turn back. Too much is caught up and too much is at stake. It would be nice if we could respond to global warming by getting rid of cars, but wheeled vehicles are so integrated into modern life that it would be impossible to run a contemporary city without them. Rather than get rid of cars, we seek other solutions, which often only increase entanglements, as seen in the enormous battery factories being built by Tesla, or in the environmental problems caused by mining for rare earths.

It might be argued that a simpler explanation for the increase in human-thing dependence is that increasing entanglements enabled humans to harness more energy and do more. I suggested in chapter 5 that dependence on things does allow humans to do more. They became agriculturalists so that they could obtain more food from a given unit of land, and thus gain more prestige, have bigger hunting rituals, and be buffered against risk. Threshing machines and combine harvesters were much more effective at harvesting and threshing crops than obsidian sickles and wooden flails. The cumulative buildup of things that can harness energy means that humans and societies have the power to achieve more and accumulate more. The increasing spread of entangled tentacles into every nook and cranny of the globe and across space and time allows humans to extract more, exploit labor more effectively, and concentrate more energy and information. Humans depend on entanglements so that they can act and build more quickly. The speed of cotton production, harvests of cereals and opium, and transport of all sorts of goods has massively increased. According to this view, the increase in entanglements results simply from greater power to respond to problems that entanglements create. The trouble with this argument is that the flows of energy do not by themselves necessarily lead to any more than a concentration of energy, a reinforcing of the status quo. The argument does not answer the central question of this book: Why do entanglements increase? Why do humans concentrate more and more energy, especially given all the negatives that such concentrations bring with them? Answering this question returns us to the tensions between dependence and dependency, between the human dependence on things and the contradictions, conflicts, and conjunctions that result.

I would argue that entanglement theory provides a nonteleological framework for understanding humans' long-term directionality toward greater entanglement and dependence on things as a means of taking more energy from the environment. It has permitted humans to create ever larger monuments and architectural achievements, the wonders of artistic production, great operas, and sporting contests. It has meant longer and healthier lives, at least for some, better communication, more travel, better education—again, at least for some. But it has also meant that we are increasingly trapped and unable to solve global-scale problems. We have become more efficient at energy capture, but doing so has depended on developing tools to capture energy that themselves require tools and labor to produce. The result has been massive human exploitation of other humans and other species, and environmental degradation. Even Henry Ford, the inventor of the Model T and of assembly-line production, seems to have understood the dangers of what he had set in progress.

He introduced schemes that promoted rural, agrarian life in the context of a modern, technological society.[10]

Disentangling from one thing often leads to greater entanglements with another. The decline in horse-drawn transport was linked to the rise of engines with horsepower but no horses. Entanglements increased dramatically as a result. The shift from gas to electric cars is creating new entanglements. All the new developments in spinning machines, opium processing, dams, and energy generation led to more entanglements with more and varied things.

Another way of conceiving the exponential rise in human-thing entanglement is in terms of the human relationship with the "external" environment. Over the course of human evolution, the expansion of entanglements has meant that all aspects of the environment have become human artifacts. Less and less outside the human can "take care of itself." The whole environment is itself an artifact needing care and manipulation. We have begun to use the term "Anthropocene" to describe the period during which human activity has been a dominant influence on the environment, but this manipulation is not new: Early humans, as I discussed in chapter 4, used fire to transform and manage the landscape.[11] As the environment increasingly became a human artifact, humans were increasingly drawn into its management, whether in terms of managing forests, building dams, providing nutrients for agricultural soils, controlling "pests," or dealing with the effects of desertification.

EVOLUTION AS ENTANGLED COCREATION

I have used the term "evolution" very rarely in this book. Darwin, in his *Origin of Species*, preferred terms like "transmutation," because evolution implied the unrolling of a predetermined plan. The word's original meaning is to "unroll," and from classical times it had been used to describe the unrolling of a scroll or of a military maneuver. It is precisely this meaning that I want to capture. The only assumption made is that humans cannot be without things. Once humans went down the evolutionary pathway of investing in tool production, there was a necessary unrolling. The dependence on things created consequences that drew humans into further dependence on things. But this happened only because humans had the intelligence to follow down the chains of consequences they had created. Beavers struggle to build and maintain their dams, but they do not follow chains of consequences as far as humans do. We might argue, as Darwin did, that increases in brain function and increases in the ability to manipulate and use tools were closely related. Natural selection favored humans with the intelligence to follow down chains of consequences and the ability to use things to fix things.

The unrolling of the dependency relationships between humans and things is the direction of human evolution. Yet within this overall directionality, we have seen enormous diversity in the way the unfolding has occurred. The specific paths have varied greatly. In each pathway, humans have exploited the affordances, material properties, and potentials of the things around them. They have struggled to fix things, to innovate, to solve problems, to deal with the contradictions and conflicts and conjunctions of human-thing entanglements. The agency of things and the agency of humans have coproduced each other.

Human evolution is heterogeneous, a cocreation of things that are biological and material and ideational and chemical and institutional. While biological evolutionary theory may help account

for parts of this evolutionary process, it cannot explain the total-
ity of it. This is why I have avoided (for the most part) the term
"evolution." Even though recent developments in biology have
created an Extended Evolutionary Synthesis that is less gene-
centered and more consistent with an entanglement perspective,
"evolution" carries with it a load that is overly biological. What is
needed is an approach that recognizes the diversity of evolutionary domains, acknowledges the tensions among them, and builds
theory from the realities and conjunctions of interdependencies.

The same interdependencies that create a general directionality also create specific pathways. Each path that is taken creates
a legacy that is material, social, ideational, and multiply entangled. It becomes difficult to go against the grain, take the road
less traveled, swim against the tide, and all the other metaphors
that express the difficulty of disentangling onto alternative pathways. But neither specific nor general directionalities result from
arbitrary goals. In the case of general directionality, HT → (TT
and TH) → greater HT is a self-catalyzing process that produces
greater human dependence on things and greater entanglement
without any goal being set. Human goals clearly do play a part
in specific pathways, but the goals themselves are created in the
context of particular entanglements. Thus Paleolithic hunter-
gatherer societies pursued such goals as achieving a good hunt
and maintaining a balance with animal spirits. In the Neolithic in
many parts of the world, humans had goals such as pleasing ances-
tors or maintaining their agricultural fields. Medieval Europeans
valued goals such as acting according to God's word or behaving
with honor on the battlefield. Industrial Britain encouraged goals
such as progress, self-betterment, or gathering wealth. All these
goals were constructed within specific historical experiences of

entanglements. They contributed to the entanglements; they were another type of thing that became laced together. But they were not some external hand that drove development in a particular direction.

Why the Question Matters

The brain that has made our success could also
be our undoing, simply not being good enough
to manage its own creations.

—CHRISTIAN DE DUVE

WHAT ARE the ethical implications of leaving the question of directionality unanswered? If our human "being" depends on things, and if the instability of things means that we are always drawn into their care, then it follows that ever since we first picked up a tool, we were destined for our current predicament. Today humans cause such effects on the world and its climate that the lives of future generations become endangered. Entanglement theory predicts that we will try to solve the problem of climate change by further fixing things, and in fact this approach dominates discussion of solutions today. We talk about making cars more fuel-efficient and shifting to alternative forms of energy that are less dependent on carbon. Especially in the West, we have tended to deal with problems by finding technological ways of changing the world rather than by changing our dependence on things. Perhaps we are approaching the limits of this way of "being."

But there are barriers to any attempt to reverse or even slow our onward search for things. The main difficulty is that, paradoxically, we do not see the "thingness" of things. We see things as objects of desire, as inert and disconnected. In chapter 4 I showed that since the eighteenth century in Europe and the United States, the rise of consumerism has led us to focus on objects in isolation from their chains of entanglements. We see things as solving problems or bringing happiness and fulfillment, and do not see so readily the suffering, poverty, and environmental damage that complex entangled strands may cause. We are encouraged not to see the filaments and tentacles that emanate from, through, and around the flows we call things.

We think we are in control of things, and we fail to see the extent to which they control us. For instance, there are many activist campaigns that deal with fair trade, organic foods, and environmentally friendly products. But these tackle only small parts of the problem. It is easy enough to criticize our headlong rush to buy the latest hybrid cars, smartphones, and TV monitors and to complain that we are hardly aware of the full chains of operation in which they are embedded. The negative impacts are often distant in space and time and difficult to evaluate. Yet how can we do otherwise? Most of us have no real alternative to driving a car to get to work. Many of us feel a duty of citizenship, which means keeping up with events, which makes an Internet and cable TV connection—and the requisite devices—seem a necessity. The lattice of interdependencies makes change very difficult.

Perhaps we can learn from societies that have taken different pathways. For example, Karl-Erik Sveiby argues that we can study the sustainable practices of Australian Aboriginal groups over tens of thousands of years in order to explore how the headlong turn to things was avoided.[1] Aboriginal land use involved very complex management and strict practices of careful land cultivation.[2] It was often assumed that European colonization occurred in a wild, empty *terra nullius*, but in fact the continent had been carefully and thoroughly farmed and managed for millennia to create a rich abundance of animal and plant life and a deep awareness of how to prevent natural disasters.[3] Many of Australia's plant species tolerate or propagate through fire, but fire could be contained through ceremony and carefully controlled burning. In historic times, overuse of the land by colonists led to massive bush fires. Aboriginal people never had to contend with such disasters.[4] Controlled fire was the key method by which they ensured ecological diversity and regulated flora and fauna. The patchwork of burned and unburned areas often created protected areas for particular food and medicinal plants. For instance, yam daisy crops were able to flourish in millions without being eaten by kangaroos. Burned areas also limited the spread of accidental fires and made seeing while traveling easier. Between the grasslands were areas of dense forest. As William Gammage has written, the continent was so carefully looked after that Europeans often referred to areas of Australia as "parks," with terraces, lawns, grottos, and stately trees—even while they refused to accept that the land had been carefully cultivated.[5] This lack of recognition of the landscape's entanglements had lasting consequences for European agriculture. European farmers saw the rich grasslands and considered them fertile ground for plant cultivation, without understanding that these landscapes were not natural and would revert back to their original state without systematic and careful use of fire. The situation has been exacerbated by the use of nonnative plants and intensive agricultural methods, which have totally depleted the soils' nutrition.

That the sustainable use of the Australian landscape persisted for tens of thousands of years does not mean there was no change. There was increasing entanglement through time, with new technologies (e.g., the boomerang), more diverse and efficient stone tools, new migrations, new forms of art, increased territoriality linked to increased population, more village formation, and more extensive trade. But all these changes happened within a sustainable set of beliefs and practices held by Aboriginal peoples.

For another example of an alternative pathway, let us look at something more familiar to archaeologists. "Heritage" is the name given to parts of the entanglements between humans and old things. The "heritage industry" deals with conservation, management, and public access to archaeological and historic sites and monuments. Not all societies find it necessary to conserve ancient sites, and even in Western societies not all monuments are equally conserved. We tend to be more concerned with classical architecture than slag heaps from mines—although even some of the latter have recently received UNESCO World Heritage status. Such is the compulsion to save the past in modern societies. In chapter 4 I described how the flow of heritage bumped into the flow of dam building in China and Egypt (as it did elsewhere in the world) with the result that massive sums were spent lifting monuments, and huge bureaucracies were set up to deal with heritage. Across the globe, development schemes in the form of new roads, pipelines, bridges, industrial estates, and housing blocks are held up or made more expensive by the archaeological work that is mandated before construction. Many of us take for granted that the past should be saved. But is it really necessary? Could we not disentangle from our own pasts? Perhaps we could learn from the traditions of Northwest Coast native groups, who see it as positive and desirable that cultural monuments return to nature, be covered with vegetation, and decay back into the ground.[6] The monuments are alive when they are buried. Denis Byrne describes non-Western modes of treatment of valued religious and historical artifacts that involve continual use and addition.[7] Could Western conservation practices change to encourage and promote burial or integration into daily life? Could we really disentangle from heritage? It is unlikely that such radical ideas could take hold in international heritage management any time soon. There is too much vested interest in making heritage pay, too much acceptance of the idea that good conservation of sites is an indication of modernity, too much adoption of the idea that sites of universal value need to be conserved for future generations, too much governmental and private investment in heritage, too much national income from cultural tourism. It is difficult to turn things around. We think we are in control of heritage, but in fact it is in control of us.

Australian Aboriginal or Northwest Coast Native notions of the environment and heritage are caught up in their own entanglements and may be inappropriate for societies with large-scale industry and institutions. Nevertheless, these alternative ideas about natural and cultural heritage may be useful in trying to think outside the box.

INEQUALITY

If human-thing entanglement brings a relentless accumulation of stuff, it forces us to address the question of inequality. In chapter 1 I discussed studies showing that the overall global increase in energy capture also entailed an increase in the amount of human-made material items, but I also noted great divergence in the amounts of material items held by different people. Supermarkets in high-income societies discard food because it does "not look good," yet in 2015 there were 800 million people in the world without enough food.[8]

What is the link between inequality and human-thing entanglement? One answer would be to say that all societies have inequality—between men and women, between people of different ages, between more successful and less successful hunters, between charismatic or authoritative leaders and a populace—and that this universal inequality gets magnified when more material stuff is around. As the amount of human-made things increases, so can the spread between those with more and those with less. The Gini index, a commonly used measure of inequality, is sensitive to the total amount of stuff (objects or income) that is available to be owned. Archaeologists have spent much time and effort explaining the rise of inequality. Perhaps once we accept the overall increase in material things, there is nothing to explain.

According to this view there is an expanding space of possibilities that humans, given their ingenuity, seek to fill. The graphs in chapter 1 show energy capture, wealth, and technological capacity increasing exponentially—gradually at first but with an ever greater upward slope. But I noted that the graphs use data from only "more socially developed" regions. Rather than focus only on the rising line, perhaps we should consider the space under the line (figure 8.1). The line still depicts a long, slow increase in the amount of human-made stuff, but below it anything can happen. Some humans choose simple, nonmaterial, and nontechnological lives. There are examples, such as the ascetic Cathars in twelfth- to fourteenth-century CE southern France, who saw all material things as sinful, or in more recent times the Amish in Pennsylvania and other US states.[9] Other groups invest massively in matter and technology. These are the ones we call "more socially developed."

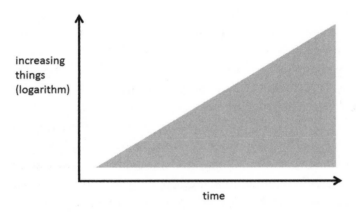

Figure 8.1. The entanglement process leads to more things through time, but inequalities increasingly emerge between those with more and less things. Source: Author.

I have argued that the upper line moves upward because of the process I have described in this book. It is a gradual process. The Hadron Collider could not have been built in the Neolithic because too much else had to happen first. The cumulative boom-and-bust process took time. And as the process evolved at the upper end, the space of possibilities below gradually expanded.

But this view takes insufficient account of the entanglements that accompany energy capture and material wealth. Increases in use of things have been associated with the forceful exploitation and domination of others. The industrial production of cotton emerged on the backs of slaves forcibly transported to the Americas, as did the production of sugar and many other commodities. The opium trade expanded alongside the military ambitions of the British and Americans and led to the addiction and decline of countless people around the world. The recovery of usable materials from discarded Christmas tree lights is shunted from the United States to China. Rare earths for electric cars and electronic devices are also mined in China. The idea of progress was used to justify the maltreatment of indigenous people around the globe under imperial expansion. In all these cases, the entanglement of things kept some people at a low level of income, often exploited and held back, while others enjoyed the benefits of increased energy capture. All entanglements can be seen as comprising intersecting operational chains or flows that get caught up in each other, and each chain involves upstream-to-downstream processes whereby dominant groups can shunt the more labor-intensive and unpleasant work onto others. Indeed the notion of an operational or supply "chain" expresses well the ways in which humans become enchained to each other. Strategic points along the chains can be controlled.

I have not focused very much on human-to-human relations in this book because there are few such relations that can be abstracted from HT and TH relations. But when we come to discuss relationships between entanglement and inequality, we have to include HH relations. Gavin Lucas and I argue that inequality is inextricably linked to human-thing entanglement.[10] Let us imagine that in a noncomplex society with few tools, I make a stone knife that is more effective than other knives at cutting and killing. I become dependent on the thing, but the thing depends on other things: the bone and the bitumen for the handle, and the pebble used to knap the flint blade—and on me to keep it sharpened. In making a new, more effective type of stone knife, I also take on a privileged position in terms of other humans. In particular I can get them to do work for me, either by force or prestige or by lending or gifting my knife or the products of its use. I can delegate to others parts of the operational sequences of procuring stone, obtaining tools, manufacturing blades, sourcing bitumen, and so on. So we can say that human dependence on things leads to thing dependence on other things, thing dependence on humans, and greater human domination over other humans.

In human-thing entanglements, some people are able to situate themselves so that they take all the benefits while others bear most of the costs. Elites tend to have more things than nonelites. This gives them more ability to control others, as well as more ability to walk away from or find solutions to problems. They have more resources available to them. As we have seen throughout this book, human-thing entanglement is enabling, and elites are enabled by the entanglements that serve them. Nonelites participate in the same entanglements, but they experience less enabling and more entrapping. Nonelites have fewer material goods that can be used to alleviate problems such as pollution, hunger, or ill health. They participate in entanglements from which they are often denied the benefits. We can talk of this as alienation, but it is a product of how people are situated in relation to entanglements. Thus Britain was able, through its control of naval power and its imperial trading connections, to enforce a global cotton-producing network centered in northern England. It was also willing to use its naval power to force an exploitative opium trade on both India and China.

Entanglement keeps some people at a low level of integration with human-made things. The wedge shape in figure 8.1 is not just the product of humans filling the space of possibilities as that space increases through time. It could have been the case that all societies saw an overall increase in the amount of material goods, a fairly even distribution of those goods through the population, and an equal participation along the threads of entanglement. But that is not what has happened. People have been forced to survive along the lower limits of the wedge, even in many cases to descend toward it. Human-thing entanglements have been an effective mechanism by which humans have exploited other humans. Few governments today promote the redistribution of wealth, and the long-term study of entanglements suggests that the more humans become dependent on things and their chains of interdependencies, the more the inequalities will increase. The violence of one human against another is closely linked to the human entanglement with things.

Contingency and Determinism

There is an ethical danger that just as network and entanglement theories disperse causality along chains, webs, filaments, and strings, they also disperse responsibility. The "buck" never "stops here." Inequality and exploitation are, according to this view, just the product of specific entanglements. In response to my description of how human exploitation could be seen as manipulating entanglements, one might counter that exploitation of humans by other humans is simply a product of particular types of entanglement. There is thus no identifiable cause of inequality. According to this view, the causes are lost in the entanglements.

I have also not resolved the question of whether the increase in human-thing entanglement is the result of contingency or is determined. On the one hand, I have talked of the contingencies and conjunctions that set human-thing dependencies down particular pathways. On the other hand, I have suggested that once a pathway has been taken, the webs of interdependencies make it difficult to change to a different path; evolution is an unrolling. So which is it? Is the pathway taken contingent, chance, or predetermined? The wider debate is best exemplified by the difference between Stephen Jay Gould—who argued that, because mutations happen by chance, if you reran the evolutionary tape you would always get a different result—and Simon Conway Morris, who argued for greater determinism.[11] In ethical terms, could human agency change or rewind the tape in order to better the conditions of others? Can we do so in the future? Or is everything determined? Is human development the result of contingency or determinism?

The answer is probably both. The biochemist Christian de Duve identified a range of evolutionary processes from determinism to constraint to contingency.[12] At one end there is deterministic necessity. Things could not have been otherwise, given the conditions that existed. Except at the subatomic level where quantum mechanics plays a role, most physical and chemical processes belong to this category. Processes follow laws of nature. As one moves away from such determinism, various forms of bottleneck provide constraints. For example, there are external selection processes, of which Darwinian natural selection is the best-known example, that may allow only one option to work out. There are also internal constraints, such as the structure of the genome or existing body plans, that funnel evolution down a particular pathway. Then there are mixtures of determinism and chance, as in the example of path dependency, where a contingent decision can no longer be changed once a path has been entered into. For example, a driver might arrive at a fork in the road and flip a coin about which path to take—but once this chance decision has been made, it is difficult to change path or turn back. At the far end of the continuum from determinism to contingency are extremely improbable events, unlikely to be repeated anywhere at any time.

De Duve argues that "evolutionary pathways may often have been close to obligatory, given certain environmental conditions, rather than contingent and unrepeatable, as received wisdom contends." The many instances of convergent evolution, where similar specializations have developed in widely separated parts of the world, support this idea. The same solution is often found time and again. But de Duve also recognizes the importance of being in the right place at the right time. Things have to come together—an accumulation of developments has to take place— before the conditions can be met for particular traits to be favored.

In this book I have shown through many examples that the development of human-thing entanglements required conjunctural events. These are seemingly unrelated, widely separated events that combine to produce unexpected effects. They are not "random," because they have specific causes and consequences. But their appearance is in no way predictable: They come about through unanticipated intersections, contradictions, and conflicts among heterogeneous processes, and they depend on the gradual accumulation of materials, skills, and information. For a conjunctural event to happen, other things have to have happened first. Different flows and pathways have to "wait for" each other before specific developments can occur. On the other hand, it is remarkable that agriculture emerged in different parts of the world at about the same time, that wheels for transport came into existence in Europe and Asia at about the same time, and that, as Ian Morris has shown, the overall increases in energy capture in more developed regions across the globe look very similar. I have argued that webs of interdependencies create pathways from which it is difficult to turn back once certain thresholds have been reached.

So human evolution is a bit of both: conjuncture and determinism. Human-thing entanglements have a heavy directionality, both specific and general, and like ocean liners or ships of state, they are difficult to turn once they have started moving. But it is always possible for agency to have its effects, for new conjunctures to arise, for new pathways to be suggested. The entanglements between humans and things can be critically assessed and evaluated, and alternatives implemented. It would be wrong to abdicate responsibility into the dispersed webs of entanglements. We can intervene, but the question is how.

WE CANNOT KEEP DOING WHAT WE HAVE ALWAYS DONE

This book has been an exercise in taking the blindingly obvious—the increase in the amount of human-made stuff in people's lives over time—and asking the simplest of questions: Why? It might be argued that the answer is also blindingly obvious. I have argued that humans as a species are distinguished by their facility at using things to solve problems. Humans, *Homo faber*, make things, so isn't it obvious that through time they have made and accumulated more things? Isn't the answer just simply that humans use things to make a better life, more wealth, more comfort, less disease, more travel, more enjoyment? Are not humans naturally competitive and acquisitive? Don't humans just use things so that they can have power over other people?

I hope I have shown in this book that an adequate answer is rather more complex than these shots from the hip. None of these more obvious answers explains *how* the overall amount of human-made stuff has increased. For a very long time, modern humans managed with very little stuff, and in some parts of the world they did so into historic times. Modern humans had evolved by at least one hundred thousand years ago, and yet the rate of increase in material stuff did not start rising rapidly until ten thousand years ago. Even up to the last few hundred years, the rate of accumulation was relatively slow. Modern humans spread into Australia by about seventy thousand years ago and developed a sustainable relationship with the material world that was extremely successful and complex but did not involve large-scale production of material goods. Off-the-cuff answers do not explain that pattern, or why it should have coexisted with the exponential growth seen elsewhere in the world.

But perhaps the most problematic aspect of the obvious answers is that they make assumptions about what it means to be human. In this book I have certainly made a universal assumption about humans—that they depend on things. But I have not argued that humans are "basically" competitive, acquisitive, or accumulative, or that they have this or that universal goal. On the contrary I have argued that all these aspects of being human derive from particular relationships between humans and things. What and who we are as humans depend on the things we have become caught up with; we have taken different pathways and forged different relationships with the environment and with each other.

This is not to say that we can blame things for environmental damage or massive inequalities. This book is not another anti-materialist tract. There is no point in blaming stuff. It is not the fault of things that we have gone down pathways that have produced global warming and mass global poverty and unfettered violence. As humans, we cannot give up all things and live immaterial lives. We are material beings. But we can explore ways in which as individuals we depend less on consumer goods, and seek social, economic, and political solutions that lessen our headlong drive toward material accumulation. And we can be more alert to the ways in which things inveigle us into their needs and attachments. It is not a matter of rejecting technological solutions and returning to a past way of life. This book has shown how difficult it is to go back and return to the past. But the book has also shown that modern technical solutions are likely only to increase entanglements and problematic conjunctions over the long term. Many of the techno-fixes that are proposed today for climate change, often termed "geoengineering" or "climate engineering," are truly frightening.[13] We cannot keep doing what we have always done—find short-term technological solutions that lock us into long-term pathways. We need to explore more carefully ways in which we can change ourselves and our contemporary addiction to things, and at the same time to evaluate critically the chains of entanglements that emanate from things but that we often are encouraged not to see. We need to struggle to disentangle, to swim against the tide, to seek alternatives, and to slow things down. We need to depend on things but also to follow sagely, thoughtfully, critically down the chains of consequences. We need to be more *sapiens* and less *faber*.

Notes

PREFACE

1. For example, see Van der Leeuw et al., "Toward an Integrated History"; Verburg et al., "Methods and Approaches"; Young et al., "Globalization of Socio-Ecological Systems."

2. The story of the biological evolution that led to modern humans has become vastly more complex over recent decades as new fossil discoveries have been made and many new hominin species identified (for an accessible summary, see Barras, "Where Did We Really Come From?"). It is often difficult to know which stone tools were made by which early *Homo*, and in any case early tools made of organic materials are unlikely to have survived. We also know more clearly now that chimps and many other animal species use tools in some form. Nevertheless, I have assumed in this book that a distinctive feature of the hominin line that led to *Homo sapiens* was the manufacture and use of tools (the first stone tools used by Australopithecines, if not earlier) and material symbols (perhaps by *Homo erectus*).

3. For example, Antal and Van den Bergh, "Macroeconomics"; Heinberg, *End of Growth*; Dietz and O'Neill, *Enough Is Enough*.

CHAPTER I

1. Morris, *Why the West Rules*.

2. Baumard et al., "Increased Affluence"; Morris, *Why the West Rules*; Morris, *Measure of Civilization*.

3. Shennan, "Demographic Continuities and Discontinuities"; Hodder, *Studies in Human-Thing Entanglement*, 147.

4. Cane, *First Footprints*.

5. Renfrew, "Symbol before Concept," 128.
6. Hodder, Çatalhöyük; Zeder, "Neolithic Macro-(R)Evolution."
7. Astruc, Tkaya, and Torchy, "De L'efficacité des Faucilles Néolithiques."
8. Maeda et al., "Narrowing the Harvest."
9. Quick, *Grain Harvesters*, 18. See also Lee, *Harvests*.
10. Rooijakkers, "Spinning Animal Fibres"; Zeder, "Neolithic Macro-(R) Evolution."
11. Mary MacVean, "For Many People, Gathering Possessions Is Just the Stuff of Life," *LA Times*, March 21, 2014.
12. Trentmann, *Empire of Things*.
13. Hodder, *Symbols in Action*.
14. http://www.worldwatch.org/node/810. Accessed January 12, 2018.
15. http://www.deptofnumbers.com/income/us/. Accessed January 14, 2018.
16. Darwin, *Origin of Species*; Laland, Odling-Smee, and Feldman, "Niche Construction"; Smith, "Cultural Niche Construction Theory."
17. Information on the environmental impact of cotton from the Huffington Post, https://www.huffingtonpost.com/mattias-wallander/t-shirt-environment_b_1643892.html. Accessed January 12, 2018. US transport sector from http://www.ussusa.org.
18. For genetically modified crops, see *The Guardian*, June 13, 2012. https://www.theguardian.com/environment/2012/jun/13/gm-crops-environment-study. Accessed January 12, 2018. For synthetic fibers, see Beckert, *Empire of Cotton*. For climate engineering, see https://www.theatlantic.com/magazine/archive/2009/07/re-engineering-the-earth/307552/ and Hamilton, *Earthmasters*.

CHAPTER 2

1. Bury, *Idea of Progress*; Tarlow, *Archaeology of Improvement*; Nisbet, *History of the Idea of Progress*.
2. Albright, "Jordan Valley."
3. Nisbet, *History of the Idea of Progress*.
4. Bergson, *Creative Evolution*, xiii. For other quotes from Bergson in these paragraphs, see pages 87, 102, 103, 104, 251.
5. White, *Evolution of Culture*, 39. For a contemporary example using similar arguments, see Judson, "Energy Expansions of Evolution." For other quotes from White in these paragraphs, see pages 33, 35, 37, 42, 80.
6. Trigger, "Alternative Archaeologies."
7. Smith, "Cultural Niche Construction Theory," 260. See also Zeder, "Domestication as a Model System"; Zeder and Smith, "Conversation on Agricul-

tural Origins," 688; Zeder and Spitzer, "New Insights into Broad Spectrum Communities."

8. For a contrary view, see Kremer, "Population Growth and Technological Change."

9. Stewart, *Evolution's Arrow.*

10. Huxley, *UNESCO*, 9, 10.

11. See Young et al., "Globalization of Socio-Ecological Systems"; Verburg et al., "Methods and Approaches to Modeling the Anthropocene"; and Van der Leeuw et al., "Toward an Integrated History to Guide the Future." For another discussion of complexity theory in archaeology, see Kohler, "Complex Systems and Archaeology."

12. Dupuy, *Le Sacrifice et l'Envie.*

13. McShea and Brandon, *Biology's First Law.*

14. For example, Shennan, *Genes, Memes, and Human History.*

15. Lineweaver, Davies, and Ruse, "What Is Complexity?"

16. Campbell, *Romantic Ethic.*

CHAPTER 3

1. Boyd and Richerson, *Culture and the Evolutionary Process*; Shennan, *Genes, Memes, and Human History.*

2. Darwin, *Origin of Species*; Dawkins, *Selfish Gene*; see also Fracchia and Lewontin, "Does Culture Evolve?"

3. Nisbet, *History of the Idea of Progress*, 173.

4. Gould, *Wonderful Life.* For the quote, see 228.

5. Heim et al., "Cope's Rule."

6. Benton, "Progress and Competition," 306, 307.

7. Fracchia and Lewontin, "Does Culture Evolve?" 515. For a discussion of the different measures of biological complexity that have been proposed, see Lineweaver, Davies, and Ruse, "What Is Complexity?"

8. For example, Huxley, *UNESCO.*

9. Benton, "Progress and Competition." For quote, see 330.

10. Bergson, *Creative Evolution*, 87.

11. Gould, *Wonderful Life*; Godinot, "Hasard et Direction."

12. Caldwell and Millen, "Studying Cumulative Cultural Evolution"; Boyd and Richerson, *Culture and the Evolutionary Process*; Shennan, *Genes, Memes, and Human History*; Bateson, "Behavioural Development"; Hinde and Fisher, "Further Observations"; Hirata, Watanabe, and Masao, "'Sweet-Potato Washing' Revisited"; Sherry and Galef, "Cultural Transmission."

13. Dawkins, *Selfish Gene.* Boyd and Richerson, *Culture and the Evolutionary*

Process; Shennan, *Genes, Memes, and Human History*; Durham, "When Culture Affects Behavior."

14. Barrett, "Material Constitution of Humanness."
15. Lewontin, *Triple Helix*, 20.
16. Bergson, *Creative Evolution*, 58.
17. Jablonka and Lamb, "Précis of Evolution."
18. Gilbert, Bosch, and Ledón-Rettig, "Eco-Evo-Devo."
19. Barad, *Meeting the Universe Halfway*.
20. Keller, "From Gene Action to Reactive Genomes." For quote, see 2428.
21. Dickins and Rahman, "Extended Evolutionary Synthesis"; Jablonka and Lamb, "Précis of Evolution"; Mesoudi et al., "Is Non-Genetic Inheritance Just a Proximate Mechanism?"; Abouheif et al., "Eco-Evo-Devo"; Gilbert, Bosch, and Ledón-Rettig, "Eco-Evo-Devo."
22. Hodder, Çatalhöyük.
23. Zhou, "Bioarchaeological Assemblages at Çatalhöyük."
24. Freeman, "Neolithic Revolution."
25. Larsen, "Agricultural Revolution."
26. Barad, *Meeting the Universe Halfway*.
27. Rosell et al., "Ecological Impact of Beavers"; Butler and Malanson, "Geomorphic Influences of Beaver Dams."
28. Collen and Gibson, "General Ecology of Beavers"; Naiman, Johnston, and Kelley, "Alteration of North American Streams."
29. Collen and Gibson, "General Ecology of Beavers," 440. See also Rosell et al., "Ecological Impact of Beavers."
30. Morgan, *American Beaver*, 95.
31. Miller, "Beaver Damage Control"; Arner, "Production of Duck Food."
32. Miller, "Beaver Damage Control."
33. Leslie, "Age of Consequences," 83.
34. Harrington, "Plundering the Three Gorges."
35. Hassan, "Aswan High Dam."
36. Mitchell, *Rule of Experts*.
37. Hassan, "Aswan High Dam," 75.
38. Bradley, *Past in Prehistoric Societies*; Lowenthal, *Past Is a Foreign Country*.
39. Hassan, "Aswan High Dam."
40. Morgan, *American Beaver*, 99.
41. Ibid., 249, 263.

CHAPTER 4

1. Laland, "Exploring Gene-Culture Interactions"; Laland and O'Brien, "Niche Construction Theory"; Odling-Smee, Laland, and Feldman, *Niche*

Construction; Rowley-Conwy and Layton, "Foraging and Farming"; Smith, "Niche Construction"; Sterelny and Watkins, "Neolithization in Southwest Asia"; Zeder and Smith, "Conversation on Agricultural Origins"; Boivin et al., "Ecological Consequences of Human Niche Construction."

2. Heidegger, *Poetry, Language, Thought.*

3. Hodder, *Entangled*; *Studies in Human-Thing Entanglement.*

4. Bergson, *Creative Evolution*, ix.

5. See Beckert, *Empire of Cotton*; Fitton, *Arkwrights*; Hills, "Hargreaves, Arkwright, and Crompton."

6. Beckert, *Empire of Cotton*, xix.

7. Tsing, *Mushroom at the End of the World*; Mitchell, *Carbon Democracy.*

8. Gowlett and Wrangham, "Earliest Fire in Africa"; Burton, *Fire*; Wrangham, *Catching Fire*; Glikson, "Fire and Human Evolution"; Roebroeks and Villa, "On the Earliest Evidence."

9. Burton, *Fire*, 10.

10. Dunbar, "Neocortex Size as a Constraint."

11. Aiello and Wheeler, "Expensive-Tissue Hypothesis"; Wrangham, *Catching Fire*, 112.

12. Wrangham, *Catching Fire*, 2.

13. Dickson, Oeggl, and Handley, "Iceman Reconsidered."

14. Glikson, "Fire and Human Evolution."

15. Minter, *Junkyard Planet.*

16. R. W. Neal, "Apple iPhone Uses More Energy Than a Refrigerator? Report Examines Environmental Impact of Global Tech Ecosystem," *International Business Times*, August 16, 2013.

17. Mills, *Cloud Begins with Coal.*

18. Stearns, *Consumerism in World History.*

19. Debord, *La Société du Spectacle*; Marcuse, *One-Dimensional Man*; Stearns, *Consumerism in World History*; Veblen, *Theory of the Leisure Class*; Campbell, *Romantic Ethic.*

CHAPTER 5

1. Heidegger, *Poetry, Language, Thought.*

2. Fuller, "Contrasting Patterns in Crop Domestication"; Maeda et al., "Narrowing the Harvest." For an alternative view, see Tzarfati et al., "Threshing Efficiency."

3. Booth, *Opium.*

4. Ibid., 105.

5. Hanes and Sanello, *Opium Wars*, 1.

6. Booth, *Opium*, 139.

7. Ibid., 257.

8. Ibid., 290–91.

9. Ibid., 343.

10. Darwin, *Origin of Species*.

11. Deleuze and Guattari, *Thousand Plateaus*.

12. Strathern, *Gender of the Gift*.

13. Nuttall, *Entanglement*, 3.

14. Thomas, *Entangled Objects*, 14, 207.

15. Martindale, "Entanglement and Tinkering."

16. Bourdieu, *Outline of a Theory of Practice*; Miller, *Material Culture and Mass Consumption*; Meskell, *Archaeologies of Materiality*; Malafouris, "Metaplasticity"; Watts, *Relational Archaeologies*; Olsen et al., *Archaeology*.

17. Latour, *ARAMIS*; "Agency"; "On Recalling ANT"; *Reassembling the Social*; Harman, "Entanglement"

18. Law, "After ANT," 3.

19. Ingold, "Bringing Things Back to Life."

20. Barad, *Meeting the Universe Halfway*. When Karen Barad speaks in this book of entanglement, she is referring to the inseparability of thought and world— an inseparability with which I agree. However, I give entanglement a more grounded and specific definition.

21. For example, Leroi-Gourhan, *L'Homme et la Matière*; Leroi-Gourhan, *Gesture and Speech*; Lemonnier, *Technological Choices*; Schiffer, *Formation Processes*.

22. The term "entanglement" is often used in discussions of global networks. A good example is Anna Tsing's work on the matsutake mushroom in her *Mushroom at the End of the World*. In this book she frequently uses the word "entanglement" to mean links that are heterogeneous, contingent, open-ended, and often contradictory. In *Facing Gaia*, Latour often uses the term "entangle" with similar meaning, but his focus remains on distributed agency rather than on "caught-up-ness" and entrapment.

CHAPTER 6

1. For the Institute of Mechanical Engineers report, see https://www.imeche.org/policy-and-press/reports/detail/global-food-waste-not-want-not. For the US report, see http://www.farms.com/ag-industry-news/u-s-165-billion-food-waste-dilemma-669.aspx,

2. https://www.washingtonpost.com/news/wonk/wp/2016/01/13/no-one-understands-baby-carrots/?utm_term=.c57b67f57092.

3. Shennan, *Genes, Memes, and Human History*; Hodder, *Studies in Human-Thing Entanglement*, 147.

4. Hodder, "Things and the Slow Neolithic."

5. Hardy, *Tess of the d'Urbervilles*, 444–45.
6. Meadowsong, "Thomas Hardy and the Machine." For quotes, see 234.
7. Hardy, *Tess of the d'Urbervilles*. For the quotes in this paragraph, see 446–48.
8. Lee, *Harvests and Harvesting*; Long, "Development of Mechanization in English Farming."
9. Lee, *Harvests and Harvesting*, 149.

CHAPTER 7

1. Carr and Wiemers, "Decline in Lifetime Earnings Mobility."
2. Gore, "Globalization," 2.
3. Hodder, *Symbols in Action*.
4. Wang et al., "Quantifying the Waddington Landscape"; Pierson, "Increasing Returns"; Mahoney, "Path Dependence."
5. David, "Clio and the Economics of QWERTY."
6. Arthur, *Increasing Returns and Path Dependence*.
7. Bulliet, *Wheel*.
8. Fuller and Rowlands, "Towards a Long-Term Macro-Geography," 23, 147.
9. Mitchell, *Carbon Democracy*, 238.
10. Mullin, "Henry Ford and Field and Factory."
11. Balter, "Archaeologists Say the 'Anthropocene' Is Here"; Glikson, "Fire and Human Evolution."

CHAPTER 8

1. Sveiby, "Aboriginal Principles for Sustainable Development"; "Collective Leadership with Power Symmetry."
2. I am grateful to Anna Fagan for references and thoughts in this section.
3. Langton, "Emerging Environmental Issues"; Gammage, *Biggest Estate on Earth*.
4. Bird et al., "'Fire Stick Farming' Hypothesis"; Bowman, Walsh, and Prior, "Landscape Analysis of Aboriginal Fire Management"; Russell-Smith et al., "Aboriginal Resource Utilization."
5. Gammage, *Biggest Estate on Earth*.
6. Jones, "'They Made It a Living Thing, Didn't They . . . ?'"
7. Byrne, "Buddhist Stupa."
8. Von Grebmer et al., *2015 Global Hunger Index;* For the quantification of increasing global inequality, see Deaton, *Great Escape*.
9. Ladurie, *Montaillou*.
10. Hodder and Lucas, "Symmetries and Asymmetries of Human-Thing Relations."

11. Morris, *Life's Solution*.
12. De Duve, *Singularities*. For the quote in the following paragraph, see 235.
13. Hamilton, *Earthmasters*, describes examples such as iron fertilizing the oceans, cloud whitening, and sulfate aerosol spraying in the upper atmosphere.

Bibliography

Abouheif, E., M. J. Favé, A. S. Ibarrarán-Viniegra, M. P. Lesoway, A. R. M. Rafiqi, and R. Rajakumar. "Eco-Evo-Devo: The Time Has Come." In *Ecological Genomics: Ecology and the Evolution of Genes and Genomes*, edited by C. R. Landry and N. Aubin-Horth, 107–25. Dordrecht: Springer Netherlands, 2014.

Aiello, L. C., and P. Wheeler. "The Expensive-Tissue Hypothesis: The Brain and the Digestive System in Human and Primate Evolution." *Current Anthropology* 36, no. 2 (1995): 199–221.

Albright, W. F. "The Jordan Valley in the Bronze Age." *Annual of the American Schools of Oriental Research* 6 (1924): 13–74.

Antal, M., and J. C. Van den Bergh. "Macroeconomics, Financial Crisis, and the Environment: Strategies for a Sustainability Transition." *Environmental Innovation and Societal Transitions* 6 (2013): 47–66.

Arner, D. H. "Production of Duck Food in Beaver Ponds." *Journal of Wildlife Management* 27, no. 1 (1963): 76–81.

Arthur, W. B. *Increasing Returns and Path Dependence in the Economy.* Ann Arbor: University of Michigan Press, 1994.

Astruc, L., M. B. Tkaya, and L. Torchy. "De L'efficacité des Faucilles Néolithiques au Proche-Orient: Approche Expérimentale." *Bulletin de la Société Préhistorique Française* 109, no. 4 (2012): 671–87.

Balter, M. "Archaeologists Say the 'Anthropocene' Is Here—But It Began Long Ago." *Science* 340, no. 6130 (2013): 261–62.

Barad, K. *Meeting the Universe Halfway: Quantum Physics and the Entanglement of Matter and Meaning.* Durham, NC: Duke University Press, 2007.

Barras, C. "Where Did We Really Come From? Untangling the New Story of Human Evolution." *New Scientist* 3140 (2017): 28–33.

Barrett, J. C. "The Material Constitution of Humanness." *Archaeological Dialogues* 21, no. 1 (2014): 65–74.

Bateson, P. P. "Behavioural Development and Evolutionary Processes." In *Current Problems in Sociobiology*, edited by King's College Sociobiology Group, 363–80. Cambridge: Cambridge University Press, 1982.

Baumard, N., A. Hyafil, I. Morris, and P. Boyer. "Increased Affluence Explains the Emergence of Ascetic Wisdoms and Moralizing Religions." *Current Biology* 25, no. 1 (2015): 10–15.

Beckert, S. *Empire of Cotton: A Global History.* New York: Vintage, 2014.

Benton, M. J. "Progress and Competition in Macroevolution." *Biological Reviews* 62, no. 3 (1987): 305–38.

Bergson, H. *Creative Evolution.* New York: Holt, 1911.

Bird, R. B., D. W. Bird, B. F. Codding, C. H. Parker, and J. H. Jones. "The 'Fire Stick Farming' Hypothesis: Australian Aboriginal Foraging Strategies, Biodiversity, and Anthropogenic Fire Mosaics." *Proceedings of the National Academy of Sciences* 105, no. 39 (2008): 14796–801.

Boivin, N. L., M. A. Zeder, D. Q. Fuller, A. Crowther, G. Larson, J. M. Erlandson, T. Denham, and M. D. Petraglia. "Ecological Consequences of Human Niche Construction: Examining Long-Term Anthropogenic Shaping of Global Species Distributions." *Proceedings of the National Academy of Sciences* 113, no. 23 (2016): 6388–96.

Booth, Martin. *Opium: A History.* New York: St. Martin's Griffin, 1996.

Bourdieu, P. *Outline of a Theory of Practice.* Cambridge: Cambridge University Press, 1977.

Bowman, D. M., A. Walsh, and L. D. Prior. "Landscape Analysis of Aboriginal Fire Management in Central Arnhem Land, North Australia." *Journal of Biogeography* 31, no. 2 (2004): 207–23.

Boyd, R., and P. J. Richerson. *Culture and the Evolutionary Process.* Chicago: University of Chicago Press, 1985.

Bradley, R. *The Past in Prehistoric Societies.* London: Routledge, 2002.

Bulliet, R. W. *The Wheel: Inventions and Reinventions.* New York: Columbia University Press, 2016.

Burton, F. D. *Fire: The Spark That Ignited Human Evolution.* Albuquerque: University of New Mexico Press, 2009.

Bury, J. B. *The Idea of Progress. An Inquiry into Its Origin and Growth.* New York: Mineola, 1987.

Butler, D. R., and G. P. Malanson. "The Geomorphic Influences of Beaver Dams and Failures of Beaver Dams." *Geomorphology* 71, no. 1 (2005): 48–60.

Byrne, D. "Buddhist Stupa and Thai Social Practice." *World Archaeology* 27, no. 2 (1995): 266–81.

Caldwell, C. A., and A. E. Millen. "Studying Cumulative Cultural Evolution in

BIBLIOGRAPHY

the Laboratory." *Philosophical Transactions of the Royal Society of London B: Biological Sciences* 363, no. 1509 (2008): 3529–39.

Campbell, C. *The Romantic Ethic and the Spirit of Modern Consumerism.* Oxford: Blackwell, 1987.

Cane, S. *First Footprints: The Epic Story of the First Australians.* Crows Nest, Australia: Allen and Unwin, 2013.

Carr, M., and E. E. Wiemers. "The Decline in Lifetime Earnings Mobility in the US: Evidence from Survey-Linked Administrative Data." Department of Economics, University of Massachusetts, Boston, 2016. http://www.solejole.org/16399.pdf.

Collen, P., and R. J. Gibson. "The General Ecology of Beavers (Castor Spp.), as Related to Their Influence on Stream Ecosystems and Riparian Habitats, and the Subsequent Effects on Fish—A Review." *Reviews in Fish Biology and Fisheries* 10, no. 4 (2000): 439–61.

Darwin, C. *The Origin of Species.* New York: Penguin, 1859.

David, P. A. "Clio and the Economics of QWERTY." *American Economic Review* 75, no. 2 (1985): 332–37.

Dawkins, R. *The Selfish Gene.* Oxford: Oxford University Press, 1976.

Deaton, A. *The Great Escape: Health, Wealth, and the Origins of Inequality.* Princeton, NJ: Princeton University Press, 2013.

Debord, G. *La Société du Spectacle.* Paris: Buchet-Chastel, 1967.

De Duve, C. *Singularities.* Cambridge: Cambridge University Press, 2005.

Deleuze, G., and F. Guattari. *A Thousand Plateaus.* Translated by B. Massumi. London: Continuum, 2004.

Descola, P. *In the Society of Nature: A Native Ecology in Amazonia.* Cambridge: Cambridge University Press, 1994.

Diamond, J. *Guns, Germs, and Steel: The Fates of Human Societies.* New York: Norton, 1997.

Dickins, T. E., and Q. Rahman. "The Extended Evolutionary Synthesis and the Role of Soft Inheritance in Evolution." *Proceedings of the Royal Society B* 279, no. 1740 (2012): 2913–21.

Dickson, J. H., K. Oeggl, and L. L. Handley. "The Iceman Reconsidered." *Scientific American* 288, no. 5 (2003): 70–79.

Dietz, R., and D. W. O'Neill. *Enough Is Enough: Building a Sustainable Economy in a World of Finite Resources.* London: Routledge. 2013.

Dunbar, R. I. "Neocortex Size as a Constraint on Group Size in Primates." *Journal of Human Evolution* 22, no. 6 (1992): 469–93.

Dupuy, J. P. *Le Sacrifice et l'Envie. Le Libéralisme aux Prises avec la Justice Sociale.* Paris: Calmann-Lévy, 1992.

Durham, W. H. "When Culture Affects Behavior: A New Look at Kuru." In

159

Explaining Culture Scientifically, edited by M. Brown, 139–61. Seattle: University of Washington Press, 2008.

Fitton, R. S. *The Arkwrights: Spinners of Fortune.* Manchester: Manchester University Press, 1989.

Fracchia, J., and R. C. Lewontin. "Does Culture Evolve?" *History and Theory* 38, no. 4 (1999): 52–78.

Freeman, H. J. "The Neolithic Revolution and Subsequent Emergence of the Celiac Affection." *International Journal of Celiac Disease* 1, no. 1 (2013): 19–22.

Fuller, D. Q. "Contrasting Patterns in Crop Domestication and Domestication Rates: Recent Archaeobotanical Insights from the Old World." *Annals of Botany* 100, no. 5 (2007): 903–24.

Fuller, D. Q., and M. Rowlands. "Toward a Long-Term Macro-Geography of Cultural Substances: Food and Sacrifice Traditions in East, West, and South Asia." *Chinese Review of Anthropology* 12 (2009): 1–37.

Fuller, D. Q., C. Stevens, L. Lucas, C. Murphy, and L. Qin. "Entanglements and Entrapment on the Pathway toward Domestication." In *The Archaeology of Entanglement*, edited by L. Der and F. Fernandini, 151–72. Walnut Creek, CA: Left Coast Press, 2016.

Gammage, W. *The Biggest Estate on Earth: How Aborigines Made Australia.* London: Allen & Unwin, 2011.

Gilbert, S. F., T. C. Bosch, and C. Ledón-Rettig. "Eco-Evo-Devo: Developmental Symbiosis and Developmental Plasticity as Evolutionary Agents." *Nature Reviews Genetics* 16, no. 10 (2015): 611–22.

Glikson, A. "Fire and Human Evolution: The Deep-Time Blueprints of the Anthropocene." *Anthropocene* 3 (2013): 89–92.

Godinot, M. "Hasard et Direction en Histoire Évolutive." *Laval Théologique et Philosophique* 61, no. 3 (2005): 497–514.

Gore, C. "Globalization, the International Poverty Trap and Chronic Poverty in the Least Developed Countries." Chronic Poverty Research Center Working Paper No. 30, United Nations—Conference on Trade and Development (UNCTAD), Geneva, 2003. http://ssrn.com/abstract=1754435.

Gould, S. J. *Wonderful Life. The Burgess Shale and the Nature of History.* London: Norton, 1989.

Gowlett, J. A., and R. W. Wrangham. "Earliest Fire in Africa: Toward the Convergence of Archaeological Evidence and the Cooking Hypothesis." *Azania: Archaeological Research in Africa* 48, no. 1 (2013): 5–30.

Hamilton, C. *Earthmasters: The Dawn of the Age of Climate Engineering.* New Haven, CT: Yale University Press, 2014.

Hanes, W. T., and F. Sanello. *The Opium Wars.* Naperville, IL: Sourcebooks, 2002.

Hardy, T. *Tess of the d'Urbervilles.* Oxford: Clarendon Press, 1892.

Harman, G. "Entanglement and Relation: A Response to Bruno Latour and Ian Hodder." *New Literary History* 45, no. 1 (2014): 37–49.

Harrington, S. "Plundering the Three Gorges." *Archaeology*, May 14, 1998. archive.archaeology.org/online/news/china.html.

Hassan, F. A. "The Aswan High Dam and the International Rescue Nubia Campaign." *African Archaeological Review* 24, no. 3–4 (2007): 73–94.

Heidegger, M. *Poetry, Language, Thought.* Translated by A. Hofstadter. London: Harper, 1971.

Heim, N. A., M. L. Knope, E. K. Schaal, S. C. Wang, and J. L. Payne. "Cope's Rule in the Evolution of Marine Animals." *Science* 347, no. 6224 (2015): 867–70.

Heinberg, R. *The End of Growth: Adapting to Our New Economic Reality.* Gabriola Island, Canada: New Society Publishers, 2011.

Hills, R. L. "Hargreaves, Arkwright, and Crompton. Why Three Inventors?" *Textile History* 10, no. 1 (1979): 114–26.

Hinde, R. A., and J. Fisher. "Further Observations on the Opening of Milk Bottles by Birds." *British Birds* 44 (1951): 393–96.

Hirata, S., K. Watanabe, and K. Masao. "'Sweet-Potato Washing' Revisited." In *Primate Origins of Human Cognition and Behavior*, edited by T. Matsuzawa, 487–508. Tokyo: Springer Japan, 2008.

Hodder, I. Çatalhöyük: *The Leopard's Tale.* London: Thames and Hudson, 2006.

——. *Entangled. An Archaeology of the Relationships between Humans and Things.* Oxford: Wiley Blackwell, 2012.

—— I. *Studies in Human-Thing Entanglement.* Stanford, CA: self-published, 2016. http://www.ian-hodder.com/books/studies-human-thing-entanglement.

——. *Symbols in Action.* Cambridge: Cambridge University Press, 1982.

——. "Things and the Slow Neolithic: The Middle Eastern Transformation." *Journal of Archaeological Method and Theory*, 2017. dOI:10.1007/s10816-017-9336-0.

Hodder, I., and G. Lucas. "The Symmetries and Asymmetries of Human-Thing Relations: A Dialogue." *Archaeological Dialogues* 25, no. 1 (2017) 119–37.

Hodder, I. and A. Mol. "Network Analysis and Entanglement," *Journal of Archaeological Method* 23, Vol. 4 (2016): 1066–1094.

Huxley, J. *UNESCO: Its Purpose and Philosophy.* London: Euston Grove Press, 1946.

Ingold, T. "Bringing Things Back to Life: Creative Entanglements in a World of Materials." ESRC National Centre for Research Methods. Working Paper Series 05/10, Realities / Morgan Centre, University of Manchester, 2010. http://eprints.ncrm.ac.uk/1306/.

Jablonka, E., and M. J. Lamb. "Précis of Evolution in Four Dimensions." *Behavioral and Brain Sciences* 30, no. 4 (2007): 353–65.

Jones, S. "'They Made It a Living Thing, Didn't They...?' The Growth of Things and the Fossilisation of Heritage." In *A Future for Archaeology: The Past in the Present*, edited by R. Layton, S. Shennan, and P. Stone, 107–22. London: UCL Press, 2006.

Judson, O. P. "The Energy Expansions of Evolution." *Nature Ecology and Evolution* 1, no. 0138 (2017). doi:10.1038/s41559-017-0138.

Keller, E. F. "From Gene Action to Reactive Genomes." *Journal of Physiology* 592, no. 11 (2014): 2423–29.

Kohler, T. A. "Complex Systems and Archaeology." In *Archaeological Theory Today*, edited by I. Hodder, 93–123. Cambridge, MA: Polity Press, 2012.

Kremer, M. "Population Growth and Technological Change: One Million BC to 1990." *Quarterly Journal of Economics* 108, no. 3 (1993): 681–716.

Ladurie, E. L. R. *Montaillou: Cathars and Catholics in a French Village, 1294–1324.* Hammondsworth: Penguin UK, 2013.

Laland, K. N. "Exploring Gene-Culture Interactions: Insights from Handedness, Sexual Selection, and Niche-Construction Case Studies." *Philosophical Transactions of the Royal Society B: Biological Sciences* 363, no. 1509 (2008): 3577–89.

Laland, K. N., and M. J. O'Brien. "Niche Construction Theory and Archaeology." *Journal of Archaeological Method and Theory* 17, no. 4 (2010): 303–22.

Laland, K. N., F. J. Odling-Smee, and M. W. Feldman. "Niche Construction, Ecological Inheritance, and Cycles of Contingency in Evolution." In *Cycles of Contingency: Developmental Systems and Evolution*, edited by S. Oyama, P. E. Griffiths, and R. D. Gray, 117–26. Cambridge, MA: MIT Press, 2001.

Langton, M. "Emerging Environmental Issues for Indigenous Peoples in Northern Australia." In *Quality of Human Resources: Gender and Indigenous Peoples*, edited by E. Barbieri-Masini, 84–111. Oxford: UNESCO-EOLSS, 2004.

Larsen, C. S. "The Agricultural Revolution as Environmental Catastrophe: Implications for Health and Lifestyle in the Holocene." *Quaternary International* 150, no. 1 (2006): 12–20.

Latour, B. "Agency at the Time of the Anthropocene." *New Literary History* 45, no. 1 (2014): 1–18.

——. *ARAMIS, or the Love for Technology.* Cambridge, MA: Harvard University Press, 1996.

——. *Facing Gaia.* Cambridge: Polity Press, 2017.

——. "On Recalling ANT." In *Actor Network Theory and After*, edited by J. Law and J. Hassard, 15–25. Oxford: Blackwell and the Sociological Review, 1999.

——. *Reassembling the Social: An Introduction to Actor-Network-Theory.* Oxford: Oxford University Press, 2005.

Law, J. "After ANT: Complexity, Naming, and Topology." In *Actor Network Theory and After*, edited by J. Law and J. Hassard, 1–14. Oxford: Blackwell and the Sociological Review, 1999.

Lee, N. E. *Harvests and Harvesting through the Ages.* Cambridge: Cambridge University Press, 1960.

Lemonnier, P., ed. *Technological Choices: Transformation in Material Cultures since the Neolithic.* London: Routledge, 1993.

Leroi-Gourhan, A. *Gesture and Speech.* Cambridge, MA: MIT Press, 1993.

———. *L'Homme et la Matière.* Paris: Albin Michel, 1943.

Leslie, J. "The Age of Consequences: A Short History of Dams." In *Water Consciousness*, edited by T. Lohan, 82–97. San Francisco: AlterNet Books, 2008.

Lewontin, R. C. *The Triple Helix: Gene, Organism, and Environment.* Cambridge, MA: Harvard University Press, 2001.

Lineweaver, C. H., P. C. W. Davies, and M. Ruse. "What Is Complexity? Is it Increasing?" In *Complexity and the Arrow of Time*, edited by Charles H. Lineweaver, Paul C. W. Davies, and Michael Ruse, 3–16. Cambridge: Cambridge University Press, 2013.

Long, W. H. "The Development of Mechanization in English Farming." *Agricultural History Review* 11, no. 1 (1963): 15–26.

Lowenthal, D. *The Past Is a Foreign Country.* Cambridge: Cambridge University Press, 1999.

Maeda, O., L. Lucas, F. Silva, K. I. Tanno, and D. Q. Fuller. "Narrowing the Harvest: Increasing Sickle Investment and the Rise of Domesticated Cereal Agriculture in the Fertile Crescent." *Quaternary Science Reviews* 145 (2016): 226–37.

Mahoney, J. "Path Dependence in Historical Sociology." *Theory and Society* 29 (2000): 507–48.

Malafouris, L. "Metaplasticity and the Primacy of Material Engagement." *Time and Mind* 8, no. 4 (2015): 351–71.

Marcuse, H. *One-Dimensional Man: Studies in the Ideology of Advanced Industrial Society.* London: Routledge, 1967.

Martindale, A. "Entanglement and Tinkering: Structural History in the Archaeology of the Northern Tsimshian." *Journal of Social Archaeology* 9 (2009): 59–91.

McShea, D. W., and R. N. Brandon. *Biology's First Law: The Tendency for Diversity and Complexity to Increase in Evolutionary Systems.* Chicago: University of Chicago Press, 2010.

Meadowsong, Z. "Thomas Hardy and the Machine: The Mechanical Deformation of Narrative Realism in *Tess of the d'Urbervilles*." *Nineteenth-Century Literature* 64, no. 2 (2009): 225–48.

Meskell, L., ed. *Archaeologies of Materiality.* London: John Wiley & Sons, 2008.

Mesoudi, A., S. Blanchet, A. Charmantier, E. Danchin, L. Fogarty, E. Jablonka, K. N. Laland, T. J. Morgan, G. B. Müller, F. J. Odling-Smee, and B. Pujol. "Is Non-Genetic Inheritance Just a Proximate Mechanism? A Corroboration of the Extended Evolutionary Synthesis." *Biological Theory* 7, no. 3 (2013): 189–95.

Miller, D. *Material Culture and Mass Consumption*. Oxford: Wiley, 1987.

Miller, J. E. "Beaver Damage Control." *Great Plains Wildlife Damage Control Workshop Proceedings* 200 (1975): 23–27. http://digitalcommons.unl.edu/gpwdcwp/200.

Mills, M. *The Cloud Begins with Coal: Big Data, Big Networks, Big Infrastructure, and Big Power—An Overview of the Electricity Used by the Global Digital Ecosystem*. National Mining Association / American Coalition for Clean Coal Electricity, 2013. https://www.tech-pundit.com/wp-content/uploads/2013/07/Cloud_Begins_With_Coal.pdf?c761ac&c761ac.

Minter, A. *Junkyard Planet*. New York: Bloomsbury Press, 2013.

Mitchell, T. *Carbon Democracy. Political Power in the Age of Oil*. New Haven, CT: Yale University Press, 2011.

——. *The Rule of Experts: Egypt, Techno-Politics, Modernity*. Berkeley: University of California Press, 2002.

Morgan, L. H. *The American Beaver and His Works*. Philadelphia: Lippincott, 1868.

Morris, I. *The Measure of Civilization: How Social Development Decides the Fate of Nations*. Princeton, NJ: Princeton University Press, 2013.

——. *Why the West Rules—For Now: The Patterns of History and What They Reveal about the Future*. London: Profile Books, 2010.

Morris, S. C. *Life's Solution: Inevitable Humans in a Lonely Universe*. Cambridge: Cambridge University Press, 2003.

Mullin, J. R. "Henry Ford and Field and Factory: An Analysis of the Ford Sponsored Village Industries Experiment in Michigan, 1918–1941." *Journal of the American Planning Association* 48, no. 4 (1982): 419–31.

Naiman, R. J., C. A. Johnston, and J. C. Kelley. "Alteration of North American Streams by Beaver." *BioScience* 38, no. 11 (1998): 753–62.

Nisbet, R. A. *History of the Idea of Progress*. New Brunswick, NJ: Transaction Publishers, 1980.

Nuttall, S. *Entanglement. Literary and Cultural Reflections on Post-Apartheid*. Johannesburg: Witwatersrand University Press, 2009.

Odling-Smee, F. J., K. N. Laland, and M. W. Feldman. *Niche Construction: The Neglected Process in Evolution*. Princeton, NJ: Princeton University Press, 2003.

Olsen, B., M. Shanks, T. Webmoor, and C. Witmore. *Archaeology: The Discipline of Things*. Berkeley: University of California Press, 2012.

Pierson, P. "Increasing Returns, Path Dependence, and the Study of Politics." *American Political Science Review* 94, no. 2 (2000): 251–67.

Quick, G. E. *The Grain Harvesters.* St. Joseph, MI: American Society of Agricultural Engineers, 1978.

Renfrew, C. "Symbol before Concept." In *Archaeological Theory Today,* edited by I. Hodder, 122–40. Cambridge: Polity Press, 2001.

Roebroeks, W., and P. Villa. "On the Earliest Evidence for Habitual Use of Fire in Europe." *Proceedings of the National Academy of Sciences* 108, no. 13 (2011): 5209–14.

Rooijakkers, C. T. "Spinning Animal Fibres at Late Neolithic Tell Sabi Abyad, Syria?" *Paléorient* 38, no. 1 (2012): 93–109.

Rosell, F., O. Bozser, P. Collen, and H. Parker. "Ecological Impact of Beavers Castor Fiber and Castor Canadensis and Their Ability to Modify Ecosystems." *Mammal Review* 35, no. 3-4 (2005): 248–76.

Rowlands, M., and D. Q. Fuller. "Moudre ou Faire Bouiller?" *Techniques & Culture* 52–53 (2009): 120–47.

Rowley-Conwy, P., and R. Layton. "Foraging and Farming as Niche Construction: Stable and Unstable Adaptations." *Philosophical Transactions of the Royal Society B: Biological Sciences* 366, no. 1566 (2011): 849–62.

Russell-Smith, J., D. Lucas, M. Gapindi, B. Gunbunuka, N. Kapirigi, G. Namingum, K. Lucas, P. Giuliani, and G. Chaloupka. "Aboriginal Resource Utilization and Fire Management Practice in Western Arnhem Land, Monsoonal Northern Australia: Notes for Prehistory, Lessons for the Future." *Human Ecology* 25, no. 2 (1997): 159–95.

Sagan, C. *Cosmos.* New York: Random House, 1996.

Schiffer, M. B. *Formation Processes of the Archaeological Record.* Albuquerque: University of New Mexico Press, 1987.

Shakespeare, William. *Hamlet.* Toronto: Oxford University Press, 1963.

Shennan, S. "Demographic Continuities and Discontinuities in Neolithic Europe: Evidence, Methods, and Implications." *Journal of Archaeological Method and Theory* 20, no. 2 (2013): 300–311.

———. *Genes, Memes, and Human History: Darwinian Archaeology and Cultural Evolution.* London: Thames and Hudson, 2002.

Sherry, D. F., and B. G. Galef. "Cultural Transmission without Imitation—Milk Bottle Opening by Birds." *Animal Behaviour* 32 (1984): 937–38.

Smith, B. D. "A Cultural Niche Construction Theory of Initial Domestication." *Biological Theory* 6, no. 3 (2012): 260–71.

———. "Niche Construction and the Behavioral Context of Plant and Animal Domestication." *Evolutionary Anthropology: Issues, News, and Reviews* 16, no. 5 (2007): 188–99.

Stearns, P. N. *Consumerism in World History: The Global Transformation of Desire.* London: Routledge, 2006.

Sterelny, K., and T. Watkins. "Neolithization in Southwest Asia in a Context of Niche Construction Theory." *Cambridge Archaeological Journal* 25, no. 3 (2015): 673–91.

Stewart, J. *Evolution's Arrow: The Direction of Evolution and the Future of Humanity*. Canberra: Chapman Press, 2000.

Strathern, M. *The Gender of the Gift*. Berkeley: University of California Press, 1988.

Sveiby, K. E. "Aboriginal Principles for Sustainable Development as Told in Traditional Law Stories." *Sustainable Development* 17, no. 6 (2009): 341–56.

——. "Collective Leadership with Power Symmetry: Lessons from Aboriginal Prehistory." *Leadership* 7, no. 4 (2011): 385–414.

Tarlow, S. *The Archaeology of Improvement in Britain, 1750–1850*. Cambridge: Cambridge University Press, 2007.

Thomas, N. *Entangled Objects. Exchange, Material Culture, and Colonialism in the Pacific*. Cambridge, MA: Harvard University Press, 1991.

Trentmann, F. *Empire of Things. How We Became a World of Consumers, from the Fifteenth Century to the Twenty-First*. New York: Harper, 2016.

Trigger, B. G. "Alternative Archaeologies: Nationalist, Colonialist, Imperialist." *Man* 19 (1984): 355–70.

Tsing, A. *The Mushroom at the End of the World*. Princeton, NJ: Princeton University Press, 2015.

Tzarfati, R., Y. Saranga, V. Barak, A. Gopher, A. B. Korol, and S. Abbo. "Threshing Efficiency as an Incentive for Rapid Domestication of Emmer Wheat." *Annals of Botany* 112, no. 5 (2013): 829–37.

Unger-Hamilton, R. "The Epi-Palaeolithic Southern Levant and the Origins of Cultivation." *Current Anthropology* 30, no. 1 (1989): 88–103.

Van der Leeuw, S., R. Costanza, S. Aulenbach, S. Brewer, M. Burek, S. Cornell, C. Crumley, J. Dearing, C. Downy, L. Graumlich, and S. Heckbert. "Toward an Integrated History to Guide the Future." *Ecology and Society* 16, no. 4, part 2 (2011): 10.5751/ES-04341-160402.

Veblen, T. *The Theory of the Leisure Class: An Economic Study of Institutions*. New York: Macmillan, 1899.

Verburg, P. H., J. A. Dearing, J. G. Dyke, S. Van der Leeuw, S. Seitzinger, W. Steffen, and J. Syvitski. "Methods and Approaches to Modeling the Anthropocene." *Global Environmental Change* 39 (2016): 328–40.

Von Grebmer, K., J. Bernstein, A. de Waal, N. Prasai, S. Yin, and Y. Yohannes. *2015 Global Hunger Index: Armed Conflict and the Challenge of Hunger*. Bonn: International Food Policy Research Institute, 2015.

Wang, J., K. Zhang, L. Xu, and E. Wang. "Quantifying the Waddington Landscape and Biological Paths for Development and Differentiation." *Proceedings of the National Academy of Sciences* 108, no. 20 (2011): 8257–62.

Watts, C., ed. *Relational Archaeologies: Humans, Animals, Things*. London: Routledge, 2014.

White, L. A. *The Evolution of Culture*. New York: McGraw Hill, 1959.

Wrangham, R. *Catching Fire: How Cooking Made Us Human*. New York: Basic Books, 2009.

Young, O. R., F. Berkhout, G. C. Gallopin, M. A. Janssen, E. Ostrom, and S. Van der Leeuw. "The Globalization of Socio-Ecological Systems: An Agenda for Scientific Research." *Global Environmental Change* 16, no. 3 (2006): 304–16.

Zeder, M. A. "Domestication as a Model System for Niche Construction Theory." *Evolutionary Ecology* 30, no. 2 (2016): 325–48.

———. "The Neolithic Macro-(R)Evolution: Macroevolutionary Theory and the Study of Culture Change." *Journal of Archaeological Research* 17, no. 1 (2009): 1–63.

Zeder, M. A., and B. D. Smith. "A Conversation on Agricultural Origins." *Current Anthropology* 50, no. 5 (2009): 681–90.

Zeder, M. A., and M. D. Spitzer. "New Insights into Broad Spectrum Communities of the Early Holocene Near East: The Birds of Hallan Çemi." *Quaternary Science Reviews* 151 (2016): 140–59.

Zhou, B. "Bioarchaeological Assemblages at Çatalhöyük: A Relational Examination of Porotic Hyperostosis and Cribra Orbitalia Etiologies and Transmissions." BA diss., Stanford University, 2016.

Index